RAND NATIONAL DEFENSE RE

T0290556

Force Drawdowns and Demographic Diversity

Investigating the Impact of Force Reductions on the Demographic Diversity of the U.S. Military

Maria C. Lytell, Kenneth Kuhn, Abigail Haddad, Jefferson P. Marquis, Nelson Lim, Kimberly Curry Hall, Robert Stewart, Jennie W. Wenger

Prepared for the Office of the Secretary of Defense

For more information on this publication, visit www.rand.org/t/rr1008

Library of Congress Cataloging-in-Publication Data
ISBN: 978-0-8330-9149-9

Published by the RAND Corporation, Santa Monica, Calif.
© Copyright 2015 RAND Corporation
RAND® is a registered trademark.

Support RAND
Make a tax-deductible charitable contribution at
www.rand.org/giving/contribute

www.rand.org

Preface

In January 2012, the Department of Defense (DoD) announced plans for a major reduction in the size of the U.S. armed force—a force drawdown—in response to budgetary constraints and an upcoming end to combat operations in Afghanistan. Although the Navy and Air Force already reduced their forces earlier, in the mid-2000s, the Army and Marine Corps had not seen a major reduction since the 1990s, after the Cold War. With some exceptions, the services did not take demographic diversity into account in their drawdown goals and strategies in the 1990s. With another drawdown occurring, DoD asked the RAND Corporation to examine whether future reductions could have unintended negative consequences for racial/ethnic minorities and women. To examine this issue, RAND conducted a review and analysis of the demographic profile changes during the 1990s drawdown and the mid-2000s drawdowns in the Navy and Air Force, followed by an analysis of the potential impact of force reduction policy decisions on the demographic profile of the DoD workforce. This report should be of interest to policymakers and others concerned with how force management decisions not specifically geared toward diversity goals may nonetheless affect demographic diversity.

This research was sponsored by the Office of Diversity Management and Equal Opportunity (ODMEO) in the Office of the Under Secretary of Defense (Personnel and Readiness) and conducted within the Forces and Resources Policy Center of the RAND National Defense Research Institute, a federally funded research and development center sponsored by the Office of the Secretary of Defense, the

Joint Staff, the Unified Combatant Commands, the Navy, the Marine Corps, the defense agencies, and the defense Intelligence Community. For more information on the Forces and Resources Policy Center, see http://www.rand.org/nsrd/ndri/centers/frp.html or contact the director (contact information is provided on the web page).

Contents

Figures

Tables

Summary

In January 2012, the Department of Defense (DoD) announced plans for a large-scale reduction—or drawdown—of its military force. By fiscal year (FY) 2019, the U.S. Army may be at its smallest in decades, since before World War II (Alexander and Shalal, 2014). The Marine Corps also plans significant, albeit smaller, reductions in the coming years. The Navy is not expected to reduce its active-duty force in coming years because of its drawdown in the mid-2000s. The Air Force also drew down its forces in the mid-2000s, and as of FY 2014, planned further reductions through FY 2019.

The last drawdown to affect all four DoD services occurred in the 1990s, after the end of the Cold War. During that period, the military shrank by almost 37 percent, from about 2.17 million in FY 1987 to 1.37 million by FY 2000 (Rostker, 2013). To achieve reductions of this size, the services used a variety of strategies, such as cutting accessions, to meet drawdown goals related to cost, readiness, and fairness to the force. The Navy and Air Force also drew down their forces in the mid-2000s; the Navy enlisted force shrank the most (18 percent), followed by the Air Force officer corps and enlisted force (13 percent each), and the Navy officer corps (7 percent). The Navy's and Air Force's mid-2000s goals were fundamentally the same as in the 1990s, although the drawdown strategies somewhat differed from those used in the 1990s.

Despite having a variety of goals and strategies for the 1990s and mid-2000s drawdowns, the services had few, if any, explicit diversity goals or strategies related to the drawdowns. Based on our discussions with force management experts, demographic diversity is also not part of their recent drawdown goals and strategies. However, the drawdown could have unintended consequences for demographic diversity even when diversity is not part of drawdown decisionmaking. To address the issue of unintended consequences of drawdowns on diversity, the Office of Diversity Management and Equal Opportunity (ODMEO) in the Office of the Under Secretary of Defense (Personnel and Readiness) asked the RAND Corporation to analyze how force reductions could affect the demographic diversity of the DoD workforce. The Military Leadership Diversity Commission defines *demographic diversity* as "immutable differences among individuals, such as race/ethnicity, gender, and age, as well as to differences in personal background, such as religion, education level, and marital status" (2011b, p. 16). Our study focuses on gender and race/ethnicity, although we include education and other individual differences, such as education, in some analyses.

Study Questions and Approach

Three overarching questions guide the study:

1. How did the services conduct previous drawdowns, and what happened to demographic diversity of the force during those drawdowns?
2. How might the demographic diversity of the DoD workforce be affected in a future drawdown?
3. What policy options are available to DoD and the services to address a potentially negative impact of a drawdown on demographic diversity?

We used a variety of sources and methods to address these questions. To address the first question, we first reviewed published literature and news reports on the drawdowns in the 1990s and mid-2000s. Next, we interviewed over 50 subject matter experts in the Office of the Secretary of Defense (OSD) and the services to learn about goals, strategies, practices, and outcomes for drawdowns. Finally, we analyzed historical personnel data to understand demographic trends in the active-duty military during the 1990s and 2000s. Our historical analyses included an examination of demographic group differences in retention rates (i.e., cumulative continuation rates [CCRs]), controlling for demographic-group differences in workforce characteristics like rank and occupational category.

To address the second question, we constructed and analyzed several notional drawdown scenarios using personnel data on the fiscal year 2012 active-duty force. Specifically, we compared internal military population baselines (e.g., junior enlisted women in the Navy to all junior enlisted personnel in the Navy) to demonstrate the potential effects of different drawdown strategies on female and minority groups.

For our final task, we reinterviewed a subset of the experts in the services, namely those who work in force management policy or diversity policy, to ask about policy implications for addressing the impact of drawdowns on demographic diversity. We use their inputs and our findings from the first two tasks to offer recommendations for changes to force management policy and practices in the context of force reductions and demographic diversity.

Drawdowns of the Reserve Component and Defense Civilian Workforce

The main body of this report presents findings for the active-duty military force. A thorough analysis of active component, reserve component, and defense civilian workforce was outside the scope of the project. However, we interviewed experts about reserve and civilian drawdowns, reviewed relevant reports and news stories about those drawdowns, and present limited personnel data provided by interviewees. We present our findings for these two workforces in Appendix A (Reserve Component Drawdowns) and Appendix B (Civilian Drawdowns).

Our reserve component review reveals that the reserve component reductions have not been as severe as those for the active component. As with reductions for the active component, the services use a variety of strategies to reduce their reserve forces, with accession cuts being a dominant reduction strategy. Another trend of the 1990s reductions is that demographic diversity of the reserve forces generally increased in the 1990s. For recent reductions to the reserves, our data on demographic trends are limited. However, based on demographic data from the Army National Guard (ARNG), demographic diversity in the ARNG slightly increased between FY 2010 and July 2014. However, black and female enlisted ARNG members experienced somewhat higher rates of administrative (involuntary) separations in recent years.

Recent reductions to the defense civilian workforce are not expected to be as severe as they were in the 1990s when the civilian force shrank by 36 percent. For recent reductions, the Army may experience larger cuts to the civilian workforce than the Air Force and Navy. To reduce their civilian workforces, the services are using similar strategies as they did in the 1990s: Start with hiring freezes, voluntary separations, and early retirements; follow with involuntary separations if needed; use furloughs if absolutely necessary. Because we did not analyze demographic data on the civilian workforces, we cannot speak to broad gender and racial/ethnic trends. However, prior reviews of the 1990s civilian drawdown finds that it led to an older civilian workforce and may have disproportionately affected blue-collar and clerical workers and, possibly, women and minorities. Limited data on recent Army reductions suggest that black men and women were disproportionately affected by the Army's recent (FY 2011—March 2014) administrative separations. For their parts, the Navy does not expect a demographic impact of its small civilian reductions, and the Air Force expressed concerns about mostly voluntary separations of women, Hispanics, and persons with disabilities.

In General, Active-Duty Force Reductions in the 1990s and Mid-2000s Did Not Decrease Demographic Diversity

Increased Demographic Diversity in the 1990s

Despite major reductions in the size of the active-duty force in all four services in the 1990s, demographic diversity increased. As Figures

Figure S.1
Female Representation in the Active-Duty Military, FY 1990–2001

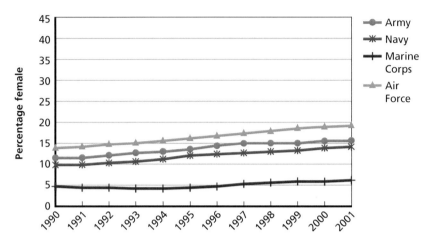

SOURCE: Analysis of enlisted and commissioned officer data from the Defense
Manpower Data Center (DMDC), FY 1990–2001.
RAND RR1008-S.1

Figure S.2
**Racial/Ethnic Minority Representation in the Active-Duty Military, FY
1990–2001**

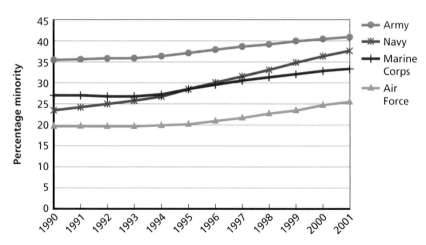

SOURCE: Analysis of enlisted and commissioned officer data from DMDC,
FY 1990–2001.
RAND RR1008-S.2

S.1 and S.2 show, the percent of females and racial/ethnic minorities increased in each service between FY 1990 and 2001. In fact, some of the larger gains occurred in the latter half of the 1990s, as the drawdown waned.

To understand these demographic trends, we decomposed the demographic changes for women, non-Hispanic blacks, and Hispanics (three major minority groups) into inflows (accessions) and outflows (separations). Focusing on female Army officers as an example, we find that female representation increased largely due to accessions, as represented by the blue bars above zero in Figure S.3. In contrast, female Army officers had relatively more separations than male Army officers; thus, change due to separations (gray bars in figure) dampened the increases in female representation (black bars) that occurred for most of the 1990s.

We also reviewed trends for Army officers by race/ethnicity and found that, in general, black representation increased from lower separations, and Hispanic representation increased from a balance of higher accessions and lower separation. However, these general trends

Figure S.3
Decomposition of Female Representation Changes, FY 1990–2001: Army Officers

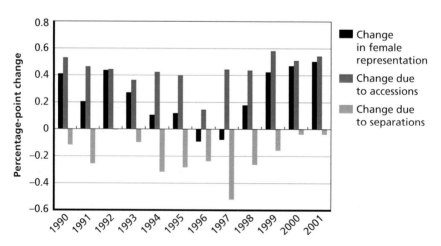

SOURCE: Analysis of DMDC data on active-duty Army personnel (FY 1990–2001).
RAND RR1008-S.3

vary to some degree when comparing different race/ethnicity groups by gender. For example, black female officers benefited relatively more from accession gains than from lower shares of separations, whereas black male officers would not have made gains by the end of the decade were it not for lower shares of separations compared to other groups.

Because women in general have higher separation (i.e., lower retention) than men, we compared male and female CCRs for the Army officer corps. As shown in Figure S.4, our adjustments to female CCRs, as shown by dashed lines, do not line up with the male observed lines. Thus, our attempt to account for gender differences in workforce characteristics like grade, education level, and years of service (YOS) do not fully explain the gender retention gap. The gap is widest between three and eight YOS both during and after the drawdown, but narrows after 11 YOS until the gap disappears around 18 YOS during drawdown and after 20 YOS postdrawdown. The findings suggest that for Army officers in the early part of their careers, gender differences in retention are not strongly related to workforce characteristics. For

Figure S.4
Actual and Adjusted Cumulative Continuation Rates, by Gender, During and After the 1990s Drawdown: Army Officers

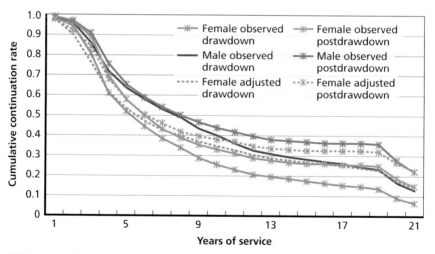

SOURCE: Analysis of DMDC data on active-duty Army officers (FY 1990–2001).
RAND RR1008-S.4

officers nearing retirement (15 or more YOS), workforce characteristics explain the gender gap.

We also compared women in different race/ethnicity groups (non-Hispanic white, non-Hispanic black, and Hispanic) to determine whether one or more groups were driving the lower female retention trends. White women had lower retention rates than black women and Hispanic women during and after the 1990s drawdown. When we adjusted black women's CCRs and Hispanic women's CCRs to look like CCRs for white women during the drawdown, we found that workforce characteristics explained most of the gap between white women's CCRs and CCRs for the other two groups of women.

Slight Decrease in Demographic Diversity in Air Force Enlisted Force in 2000s

Unlike the 1990s, demographic diversity increases did not occur for the Air Force as shown in Figures S.5 and S.6. The dip is due to the Air Force enlisted force. The Navy experienced an increase in demographic

Figure S.5
Female Representation in the Active-Duty Navy and Air Force, FY 2001–2011

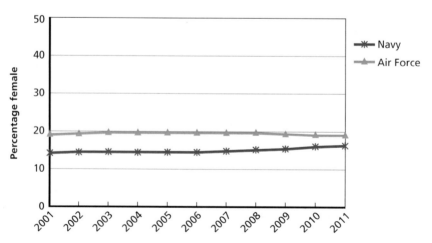

SOURCE: Analysis of enlisted and commissioned officer data from DMDC, FY 2001–2011.

RAND RR1008-S.5

Figure S.6
Racial/Ethnic Minority Representation in the Active-Duty Navy and Air Force, FY 2001–2011

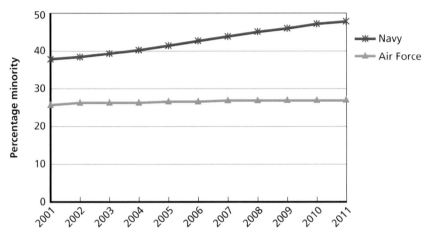

SOURCE: Analysis of enlisted and commissioned officer data from DMDC, FY 2001–2011.
RAND *RR1008-S.6*

diversity, particularly in terms of racial/ethnic diversity. Its increase in racial/ethnic minority representation was driven in large part by its enlisted force, where minority representation increased by more than 10 percentage points, from about 41 percent in FY 2001 to nearly 53 percent in FY 2011. Increased Hispanic representation was a major factor in the overall minority increase in the Navy.

As with the 1990s, we decomposed the demographic changes for the mid-2000s reductions into accessions and separations with a focus on Air Force enlisted women. As shown in Figure S.7, we find that the decrease in enlisted female representation was largely a function of lower female retention throughout the 2000s (even before the drawdown). However, the decrease was exacerbated by lower female shares of accessions in the latter half of the 2000s. Air Force accession cuts and tightening of enlisted entry standards at the beginning of the drawdown could have played a role in decreasing shares of female accessions in the latter half of the 2000s.

Figure S.7
Decomposition of Female Representation Changes, FY 2001–2011: Air Force Enlisted

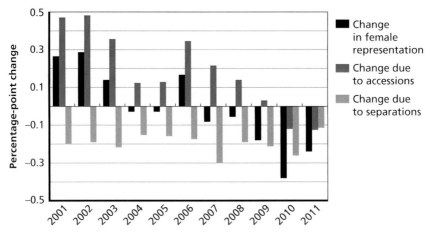

SOURCE: Analysis of DMDC data on active-duty Air Force enlisted personnel (FY 2001–2011).
RAND *RR1008-S.7*

We also examined the gender retention gap for Air Force enlisted, during and after the main drawdown period. Figure S.8 shows the results for the drawdown period. As with our analysis of the gender retention gap for Army officers in the 1990s, our adjustments to the Air Force enlisted female CCRs in the 2000s do not fully explain the gender gap in retention during and after the 2000s drawdown. Unlike the Army officer gender gap, the Air Force enlisted gender gap is narrowest around six to seven YOS, with the gap increasing in size over seven YOS. Gender differences in family responsibilities may become more salient once enlisted personnel enter their mid-20s, which occurs around six to seven YOS for most enlisted personnel.

Figure S.8
Actual and Adjusted Cumulative Continuation Rates by Gender in 2000s Drawdown Era: Air Force Enlisted

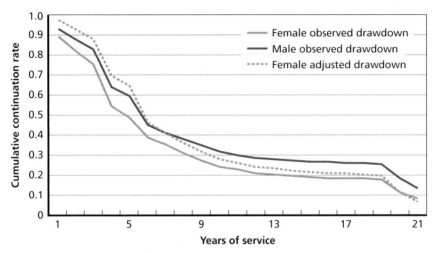

SOURCE: Analysis of DMDC data on active-duty Air Force enlisted personnel (FY 2001–2011).

RAND RR1008-S.8

Relationship Between Drawdown Strategies and Demographic Diversity

Specific Drawdown Effects on Demographic Diversity Are Unclear

Besides the goal of reducing costs, OSD and the services have a variety of goals, strategies, and practices for drawdowns. In the 1990s, Congress, OSD, and the services sought to limit involuntary separations to "keep faith" with the force. The services also aimed to reduce their forces in such a way that the right mix of skills and experience would remain to maintain readiness, and to treat personnel, particularly those asked to leave, in a fair manner. To achieve these goals, the services significantly reduced accessions, offered early retirements and voluntary separation incentives, and as a last resort, involuntarily separated personnel. The Navy and Air Force had similar goals for the mid-2000s reductions but with less emphasis on the "keep faith" goal. As a result, both services emphasized relatively more involuntary separations and relatively fewer accessions cuts than they had in the 1990s.

Because drawdowns directly target specific parts of force structure, we can assume that workforce changes directly tied to force structure are likely a result of drawdowns. For example, the 1990s drawdown resulted in an active-duty force more senior in experience, balanced more toward officers than enlisted personnel, and (at least in the officer corps) more heavily represented in nontactical operations occupations. Since heavy accession cuts reduce the junior force for years to come, accession cuts directly affect the seniority or experience of the resulting force. Likewise, cutting proportionately more enlisted personnel than officers affects the officer-enlisted balance of the force. Finally, cutting operational force structure more than infrastructure (support) affects the resulting occupational distribution of the force.

However, drawdown decisions do not have clear-cut ties to demographic diversity because the services do not make drawdown decisions with demographic goals in mind. This may explain why the literature and our interviews provided little guidance on how past drawdowns affected demographic diversity. To complicate matters, drawdowns involve a dynamic set of activities, many occurring at the same time and affecting different subpopulations in the force. Also, factors outside the services' control, such as civilian labor market opportunities and personal life choices of service members, affect demographic accession and retention trends. Teasing apart the effects of the various drawdown activities would require several details not available in the literature or provided by the services in our interviews.

Certain Drawdown Strategies Could Affect Demographic Diversity

Our review of past and current drawdown strategies and tools suggests three categories of workforce characteristics used to separate active-duty personnel in drawdowns: experience (e.g., rank, YOS, time in grade), occupational category (e.g., occupational specialties less critical to the service mission), and merit or "quality" (e.g., personnel records). Any of these could affect demographic diversity because demographic groups are distributed differently across experience, occupation, and merit categories. We therefore developed and analyzed notional drawdown scenarios to vary one or more of these workforce characteristics to examine how demographic diversity could be affected. Without all of the relevant details or personnel data to examine specific draw-down programs or tools, we gathered what details we could from our interviews and news sources to develop scenarios based on different types of drawdown programs that the services have used in the recent past (mid-2000s onward) or reportedly may use in the next few years. We tailored scenarios to each service and corps (enlisted and commissioned officer) and used FY 2012 personnel data on the active-duty force for analysis. For each scenario, we compared the proportion of cuts taken from female, black, and Hispanic groups in the target population to the proportion of cuts expected from a relevant baseline population. We assumed cuts would be randomly distributed across both target and baseline populations. These population comparisons offer a simple means of assessing the potential adverse impact on women and minorities.

We also explored whether scenario results might differ if we crossed gender and race/ethnicity groups. We selected at least one scenario from each service to analyze with the following demographic groups: non-Hispanic white women, non-Hispanic black women, Hispanic women, non-Hispanic black men, and Hispanic men. The findings from these demographic "breakout analyses" point to some boundary conditions of our main scenario analysis findings.

Table S.1 offers details on scenarios, using the Army as an example. We analyzed all of the scenarios for women, non-Hispanic black personnel, and Hispanic personnel. Scenarios with asterisks (*) were also analyzed for gender-by-race/ethnic groups (e.g., Hispanic women). The Army officer scenarios focus on reductions in force (RIFs) for captains (1a) and majors (1b), and selective early retirement boards (SERBs) for lieutenant colonels (2a) and colonels (2b). The Army enlisted scenarios include cuts to accessions while tightening Armed Forces Qualification Test (AFQT) standards (3), reductions in retention control points for E-4s (4a) and E-5s (4b), and cuts based on a Qualitative Service Program (QSP) or similar involuntary separation program for senior enlisted personnel (5).

Tables S.2 and S.3 provide results from our analysis of the Army enlisted scenarios. Cell values reflect the percentages of cuts expected for each demographic group given the scenario variation. For exam-

Table S.1
Details for Army Drawdown Scenarios

Scenario Number	Program	Targeted Population	Scenario Variation(s)	Baseline Population	Cut Sizes
1a*	RIF	Captains with 4–6 YOS	• Proxy Army Competitive Category (ACC) occupations • All occupations	All captains	• 600 • 1,200 • 1,800
1b*	RIF	Majors with 9–13 YOS	See cell above	All majors	• 200 • 550 • 800
2a	SERB	Lieutenant colonels with ≥ 4 years time in grade (TIG)	• ≥ 4 years TIG in Proxy ACC occupations • Any TIG in Proxy ACC occupations	All lieutenant colonels	• 30% of targeted population (1,953)
2b	SERB	Colonels with ≥ 4 years TIG	See cell above	All colonels	• 30% of targeted population (783)
3	Accessions cut and AFQT requirements tightened	Accessions (0–1 YOS)	• All cuts from Categories IIIB–IV (100%) • 75% Categories IIIB–IV, 25% Category IIIA • 50% Categories IIIB–IV, 50% Category IIIA	All accessions	• 10% accessions (5,185) • 20% accessions (10,371) • 30% accessions (15,556)
4a	Reduced Retention Control Points	E-4s with ≥ 8 YOS	• ≥ 8 YOS in tactical • ≥ 8 YOS in non-tactical • Any YOS in tactical • Any YOS in nontactical	All E-4s	N/A (results based on overall population proportions)

Table S.1—Continued

Scenario Number	Program	Targeted Population	Scenario Variation(s)	Baseline Population	Cut Sizes
4b	Reduced Retention Control Points	E-5s with ≥ 14 YOS	• ≥ 14 YOS in tactical • ≥ 14 YOS in nontactical • Any YOS in tactical • Any YOS in nontactical	All E-5s	N/A (results based on overall population proportions)
5*	QSP or similar involuntary separation program	E-7s, E-8s, and E-9s	• 75% of cuts from tactical, 25% from nontactical • 50% tactical, 50% nontactical • 25% tactical, 75% nontactical	All E-7s, E-8s, and E-9s	• 600 • 1,000

NOTE: *Scenarios selected for race/ethnicity breakouts by gender.

ple, the first cell in Table S.2 shows 18.8 percent for women. This means that 18.8 percent of female Army enlisted accessions could be cut if all accession cuts are taken from the lowest AFQT categories, IIIB-IV. This percentage is higher than the women's baseline of 14.7 percent, which represents the percentage of cuts to female accessions that might be expected if accessions cuts are taken across the board (not with respect to AFQT). When a scenario produces a higher percentage than the baseline, there is potential for adverse impact. Dark gray cells in the tables represent potential for adverse impact. Light gray cells reflect the opposite—i.e., less potential for adverse impact.

The accession cut scenario (3) in Table S.2 suggests that accession cuts focused on AFQT could adversely affect women, black, and Hispanic groups. The other scenarios in Table S.2 show that cuts heavily focused on nontactical operations occupations could adversely affect women, black personnel, and, to a lesser extent, Hispanic personnel. Black enlisted personnel face especially adverse effects if cuts are based on the combination of longer service and nontactical operations occupations. In contrast, cuts to personnel with longer service in tactical

Table S.2
Main Results for Army Enlisted Scenarios

Scenario Number	Scenario Variations	Women	Non-Hispanic Black	Hispanic
3	100% cuts from Categories IIIB-IV	18.8	31.7	18.0
	75% Categories IIIB-IV, 25% Category IIIA	17.8	28.8	16.9
	50% Categories IIIB-IV, 50% Category IIIA	16.7	25.8	15.7
	All accessions (baseline)	14.7	20.2	13.2
4a	≥8 YOS and in any occupational group	11.0	25.0	12.9
	≥8 YOS and in nontactical group	12.4	26.8	12.7
	≥8 YOS and in tactical group	0.5	12.1	13.8
	Any YOS and in nontactical group	18.5	23.0	12.1
	Any YOS and in tactical group	1.1	8.1	11.8
	All E-4s (baseline)	14.2	19.3	12.8
4b	≥14 YOS and in any occupational group	9.4	34.5	11.0
	≥14 YOS and in nontactical group	10.0	35.7	10.8
	≥14 YOS and in tactical group	1.1	17.5	13.9
	Any YOS and in nontactical group	15.5	23.9	14.1
	Any YOS and in tactical group	0.8	8.3	12.4
	All E-5s (baseline)	12.2	20.3	13.7
5	75% tactical, 25% nontactical	4.0	21.6	12.5
	50% tactical, 50% nontactical	7.8	26.3	12.6
	25% tactical, 75% nontactical	11.6	31.0	12.7
	All E-7s, E-8s, and E-9s (baseline)	10.9	30.1	12.7

NOTES: Except for Scenario 4, the table values provide the percentage of cuts from each demographic group for the given scenario. Scenario 3 assumes 20 percent of accessions cut. Scenario 4 is not based on specific cut sizes but reflects the demographic group proportions in the targeted and baseline populations. Scenario 5 uses a cut of 600 personnel.

Table S.3
Results for Army Enlisted Scenario with Gender-by-Race/Ethnicity
Breakouts

Scenario Number	Scenario Variations	NH White Women	NH Black Men	NH Black Women	Hispanic Men	Hispanic Women
5	75% tactical, 25% nontactical	1.0	19.3	2.3	6.8	0.3
	50% tactical, 50% nontactical	1.9	21.8	4.5	6.7	0.6
	25% tactical, 75% nontactical	2.8	24.3	6.7	6.5	0.8
	All E-7s, E-8s, and E-9s (baseline)	2.7	23.8	6.3	6.6	0.8

NOTES: Table values provide the percentage of cuts from each demographic group for the given scenario. Scenario 5 uses largest cut sizes in Table 5.3 in Chapter Five (i.e., 1,000). NH stands for "Non-Hispanic."

operations occupations could adversely affect Hispanic personnel, although cuts to Hispanic personnel without regard to service length could be adverse if instead focused on nontactical operations occupations. Table S.3 highlights the differences between Hispanic men and women in a senior enlisted reduction scenario (5). Specifically, Hispanic men could be adversely affected by cuts that lean more toward tactical operations occupations, but Hispanic women might not be adversely affected by such cuts.

Based on our analysis of scenarios across all four services, we identify three policy-relevant themes related to demographic diversity. First, cuts drawn heavily from personnel in nontactical operational occupations could adversely affect women and non-Hispanic blacks because women and non-Hispanic blacks are more heavily concentrated in nontactical operational career fields. However, Hispanic men may be adversely affected by cuts to tactical operational occupations. Second, cuts based on personnel experience could have different demographic impacts. In general, cuts based on longer service can have an adverse impact on black personnel, but cuts based on less service could adversely affect women. Third, and perhaps the strongest of the three themes, is that tightening accessions standards could have an adverse impact on women and minorities, although non-Hispanic white women are less affected by AFQT restrictions than members of other minority groups. We examined a scenario of cuts to enlisted accessions with lower scores on the AFQT. Across all four services, AFQT-based cuts show potential adverse impact against female, black, and Hispanic enlisted recruits.

Recommendations

We spoke to force management policy experts and diversity policy experts about policy options to address demographic diversity in a drawdown. These experts noted the legal challenges of using demographic information in employment decisions such as those made during drawdowns. As a result, none of the experts claimed to examine potential demographic impacts of drawdown decisions. Instead, they cited other aspects of the military career life cycle (particularly recruiting and accessions) and providing flexible career options to personnel as ways to address demographic diversity.

Because of the legal restrictions noted by the experts, we do not recommend specific changes to force management policy that would require the services to make drawdown decisions based on a person's gender, race, ethnicity, or other protected status. However, we make two related recommendations regarding how force management can consider demographic implications of drawdown decisions.

OSD(P&R) Directs the Services to Conduct Predecisional Analyses

OSD (Personnel and Readiness) (OSD[P&R]) should direct the services to conduct adverse impact analyses *prior to* making drawdown decisions, not after decisions are made. Per the *Uniform Guidelines on Employee Selection Procedures* (Equal Employment Opportunity Commission [EEOC], 1978), employers should consider the adverse impact of their employment decisions in advance of making those decisions. If adverse impact evidence is found, the next step is to validate the drawdown procedure and show "good faith" efforts to address adverse impact. Although the EEOC's uniform guidelines are not designated for military personnel management, the guidelines offer a useful approach to ensuring employment decisions are based on validated methods and measures. In the context of force drawdowns, validation questions to address include

- Is there evidence that the measures of a person's merit are empirically valid? How about the validity of the weights placed on the different measures in making drawdown selection decisions?
- Are there other measures (or combinations thereof) that are valid but have less adverse impact?
- Can the occupations, experience levels, or other factors used to target personnel be directly tied to mission requirements? Are those "requirements" valid?
- What policies or laws would have to change to address adverse impact?

To assist the services in conducting these analyses, Military Personnel Policy (MPP) within OSD (P&R) should develop policy to guide the services' efforts. ODMEO should assist in the development of this policy guidance because of its expertise in (civilian) equal employment opportunity where adverse impact analysis is common practice. The guidance should (1) list the types of questions that the analyses should address, (2) outline a general approach to analysis, and (3) require that the services briefly describe the measures and data elements used in their analyses. The general approach used in this study offers a way to structure the main analytic elements. The services should be encouraged to adjust their analyses with more detailed scenarios and modeling. The overall goal of the OSD policy is to ensure that the services "do their homework" by conducting analyses and providing appropriate documentation.

ODMEO Validates Services' Results, and OSD Directs DMDC to Acquire More Data

Given its expertise in adverse impact and related concepts, ODMEO should be responsible for validating at least some of the services' predecisional analytic results. MPP and ODMEO can specify conditions under which validation checks would not be conducted. At a minimum, ODMEO can make sure the services answered the appropriate questions stated in the policy guidance and spot-check analytical results. To spot-check results, ODMEO would need DMDC data with performance data and other details not currently available. OSD(P&R) should direct DMDC to acquire these data from the services (and direct the services to provide the data) at a level of specificity commensurate with that used by the services.

Acknowledgments

We wish to thank several people for their support of this project. We begin by thanking Clarence Johnson (Principal Director, ODMEO) for his sponsorship. We also thank the project's action officer, Bea Bernfeld (Director, Equal Employment Opportunity, ODMEO), for her continued support throughout the project. We further recognize ODMEO staff for providing helpful tips and, more generally, their expertise on diversity and inclusion.

This project could not have been completed without the generosity and insights of study participants. We thank the force management, diversity, and other policy experts in OSD and the services who described goals, strategies, and policies for accomplishing force reductions and discussed policy options for demographic diversity and force reductions. We cannot list all of their names here but wholeheartedly thank them.

We would be remiss if we did not thank our RAND colleagues for their support. John Winkler and Jennifer Lewis provided leadership throughout the project. Mady Segal and Jim Hosek reviewed the report and offered helpful guidance on how to make it better. Tara Terry, Al Robbert, and Grant Wilder offered methodological guidance, cleaned personnel data, and ran initial analyses that guided future project efforts. Andrew Madler managed the large master data sets. Miriam Matthews conducted an initial literature search that helped with our overall literature review efforts. Amy Grace Donahue took notes during discussions with subject matter experts. Roberta Shanman conducted literature searches, and Theresa DiMaggio helped organize the materials from those searches, as well as providing general administrative support throughout the project.

Abbreviations

ACC	Army Competitive Category
AFQT	Armed Forces Qualification Test
ARC	Air Force Reserve Component
ARNG	Army National Guard
ASVAB	Armed Services Vocational Aptitude Battery
AVF	All-Volunteer Force
BBA	Bipartisan Budget Act
BRAC	Base Realignment and Closure
CBO	Congressional Budget Office
CCR	cumulative continuation rate
CJR	career job reservation
DMDC	Defense Manpower Data Center
DoD	Department of Defense
DOPMA	Defense Officer Personnel Management Act
DOS	date of separation

EEOC	Equal Employment Opportunity Commission
ERB	Enlisted Retention Board
E-SERB	enhanced selective early retirement board
FY	fiscal year
GAO	General Accounting Office or Government Accountability Office
HYT	High Year Tenure
MLDC	Military Leadership Diversity Commission
MOS	military occupation specialty
MPP	Military Personnel Policy
NCO	noncommissioned officer
NDAA	National Defense Authorization Act
NH	non-Hispanic
ODMEO	Office of Diversity Management and Equal Opportunity
OPM	Office of Personnel Management
OSD	Office of the Secretary of Defense
OSD(P&R)	Office of the Secretary of Defense (Personnel and Readiness)
PBD	Program Budget Decision
PFT	physical fitness test
PPP	Priority Placement Program
QDR	Quadrennial Defense Review
QSP	Qualitative Service Program

RC	reserve component
RIF	reduction in force
RRF	retention recommendation form
SERB	selective early retirement board
SSB	Selective Separation Benefit
TERA	Temporary Early Retirement Authority
TIG	time in grade
USAR	U.S. Army Reserve
VERA	Voluntary Early Retirement Program
VRI	Voluntary Retirement Incentive
VSI	Voluntary Separation Incentive
VSIP	Voluntary Separation Incentive Pay
VSP	Voluntary Separation Pay
YOS	years of service

CHAPTER ONE

Introduction

The U.S. military, particularly the Army and Marine Corps, has begun a significant drawdown of its total force. A large portion of the drawdown will come from the active-duty force, with the Army planning to drop from force levels (as of March 2014) of 510,000 to 490,000 by fiscal year (FY) 2015, and possibly down to 420,000 by FY 2019. These proposed reductions could yield the smallest Army since before World War II (Alexander and Shalal, 2014). Although its reductions are expected to be smaller, the Marine Corps plans to make significant reductions to its workforce in the coming years. The Air Force reduced its active-duty force as recently as FY 2015.

The last major drawdown period for all services occurred between the late 1980s and the mid-1990s and is sometimes referred to as the "Post–Cold War drawdown." During that drawdown, the military shrank by almost 37 percent, from about 2.17 million in FY 1987 to 1.37 million by FY 2000 (Rostker, 2013). The Air Force took the largest cut, a 42-percent reduction, followed by Army (39 percent) and Navy (37 percent). The Marine Corps shrank by only 14 percent (Rostker, 2013). After this drawdown period ended, the next major reductions began in the mid-2000s but affected only the Navy and Air Force. The Navy's enlisted force shrank the most (18 percent), followed by Air Force enlisted (14 percent), Air Force officers (13 percent), and Navy officers (about 7 percent).

During major drawdown periods, the services balance several objectives that include reducing the budget, ensuring fair treatment to the current force, and retaining the right mix of skills and grades (i.e., avoiding a "hollow force"). Another objective that may not be explicitly considered is retaining a diverse workforce. Department of Defense (DoD) policy regarding equal opportunity and diversity management describes diversity as a "potential force multiplier in DoD mission accomplishment" (DoD Directive 1020.02, 2009, p. 3). Accordingly, DoD and the services have undertaken several efforts to achieve greater diversity, such as increasing the representation of historically underrepresented racial/ethnic groups and women in the workforce. However, large-scale personnel reductions that occur during major drawdown periods may inadvertently affect the demographic diversity of the DoD workforce. For example, drawdown strategies may target certain occupational groups that have a relatively high percentage of women or racial/ethnic minorities relative to other occupational groups. Targeting such occupational groups—all else being equal—could potentially undermine DoD efforts to achieve a force that demographically represents the nation it serves.

Defining Demographic Diversity

The Office of Diversity Management and Equal Opportunity (ODMEO) in the Office of the Under Secretary of Defense (Personnel and Readiness) asked RAND to identify and analyze potential consequences of force reductions on the demographic diversity of the DoD workforce and provide policy recommendations to address potential impacts on diversity. To conduct this study, we first need to define "demographic diversity." According to the Military Leadership Diversity Commission (MLDC), demographic diversity refers to "immutable differences among individuals, such as race/ethnicity, gender, and age, as well as to differences in personal background, such as religion, education level, and marital status" (MLDC 2011b, p. 16). In our study, we focus on gender and race/ethnicity, although we include other individual differences, such as education, in some analyses.

Demographic diversity goes beyond the mere description of individual differences in a group, such as a military workforce. Demographic diversity involves a comparison of the demographic mix of a focal group to the demographic mix of one or more baseline groups. The type of baseline group selected affects the conclusions about the focal group's demographic diversity. If the focal group is a part of the military workforce (e.g., Army active-duty enlisted force), the baseline could be the general U.S. population, the portion of the U.S. population eligible to serve in the military, or one or more internal military workforce populations (e.g., all Army active-duty personnel). Because the U.S. military has age, education, citizenship, mental, physical, and medical requirements for entry, the portion of the U.S. population eligible to serve tends to have relatively fewer women and racial/ethnic minorities than the general U.S. population. In its report to Congress, the MLDC discusses the issue of the "eligible population" as a baseline. The MLDC uses Marine Corps data to demonstrate how military entry requirements result in an eligible population less racially/ethnically diverse, and for the enlisted eligible population, less gender diverse than the general U.S. population of 17- to 29-year-olds. The implication is that a military workforce compared to the general U.S. population will be deemed less demographically diverse than one compared to the eligible population.

In our study, we use internal workforce baselines. For example, we compare the demographic composition of a particular military population (e.g., Army active-duty personnel) at an earlier time point to the composition of that workforce at a later time point. The baseline in this case would be the workforce at the earlier time point. In other cases, we compare demographic groups (e.g., men and women) within a military subpopulation (e.g., Army majors) at one point of time to demonstrate differential impacts of different drawdown strategies. Although this is not stated explicitly in the remainder of the report, we assume that increases in the proportions of women and racial/ethnic minorities in a workforce reflect higher amounts of demographic diversity in that workforce. We recognize that increases in female and minority representation could theoretically reach a point where the current majority group (namely, white men) could become a numerical minority in the workforce.

Study Questions and Approach

The study addresses three main research questions:

1. How did the services conduct previous drawdowns, and what happened to demographic diversity of the force during those drawdowns?
2. How might the demographic diversity of the DoD workforce be affected in a future drawdown?
3. What policy options are available to DoD to address a potentially negative impact of a drawdown on demographic diversity?

To address these questions, RAND performed five research tasks.

1. Review current law and policy on DoD force reductions.
2. Review the drawdown of the 1990s to identify potential effects it had on the demographic profile of the military workforce.
3. Develop and analyze scenarios of potential force reduction policies and their impact on demographic diversity of the military workforce.
4. Identify policy options for addressing potential impacts of drawdowns on diversity.
5. Develop conclusions and recommendations for policy changes that align with policy options evaluated in task 4.

The first two tasks guided tasks 3 and 4. For task 1, we held discussions with subject matter experts on current drawdown law, policies, goals, and strategies. Most of these experts have leadership roles in manpower and personnel policy organizations within the services or the Office of the Secretary of Defense (OSD). To supplement the discussion material, we reviewed policy and law about current drawdown programs and practices, as well as relevant news stories and publicly available information posted by the services about drawdown activities. For task 2, we held discussions with experts on laws, policies, goals, strategies, and outcomes for the 1990s drawdown and mid-2000s drawdowns in the Navy and Air Force. Many of these experts also provided information on current drawdown policies and practices for task

1. We reviewed literature on the 1990s drawdown to provide additional context for the discussions. We also analyzed historical personnel data to describe demographic trends during and after the drawdown and determine whether retention patterns varied by demographic group during and after the main drawdown periods.

For task 3, we developed and analyzed notional drawdown scenarios to investigate how different drawdown decisions could affect female personnel, non-Hispanic black personnel, and Hispanic personnel in the active-duty force. We supplement our main scenario analyses by examining how the intersection of gender and race/ethnicity (e.g., Hispanic women) could affect results for a select number of scenarios. Our scenario analyses used FY 2012 personnel data from the Defense Manpower Data Center (DMDC) to examine cuts to baseline populations versus populations targeted by the notional drawdown program in the scenario. We compared the proportion of cuts from female, black, and Hispanic target populations to those from baseline populations to provide a simple means of examining the potential for adverse impact against women, blacks, or Hispanics. Adverse impact is defined as "a substantially different rate of selection in hiring, promotion, or other employment decision which works to the disadvantage of members of a race, sex, or ethnic group" (Equal Employment Opportunity Commission [EEOC], 1978). We do not use the traditional "four-fifths rule" to establish evidence of adverse impact in this study but instead identify how drawdown decisions could potentially result in adverse impact based on demographic group differences in cuts to targeted and baseline populations.[1] To understand the policy implications for findings from this type of analysis, we met with force management policy experts and diversity policy experts to discuss policy options to address potential adverse impact (task 4). Based on our findings from

---

[1]   The *Uniform Guidelines on Employee Selection Procedures* (EEOC, 1978) outlined the "four-fifths rule" as a means to determine if adverse impact is present in an employment selection system. The rule compares selection ratios (i.e., number of people hired out of number of applicants) of a minority group and a majority group. If the selection ratio of a minority group (e.g., Hispanic men) is less than four-fifths of the selection ratio of a majority group (e.g., white men), adverse impact is said to exist.

tasks 1–4, we developed two related recommendations for how DoD can address policy for demographic diversity during a drawdown.

Due to time and resource constraints, the study did not perform all five tasks for all three main segments of the DoD workforce: active component, reserve component, and defense civilians. The study focused on the active component. For the other two workforce segments, we reviewed past and current drawdown goals, strategies, and practices and identified (where possible) demographic impacts of prior reductions (tasks 1 and 2). Below, we provide a brief overview of the main findings from our review of the reserve component and defense civilian drawdowns. The complete findings from our review of the reserve component and defense civilian drawdowns are in Appendix A and Appendix B, respectively.

Key Themes from Reserve Component and Defense Civilian Drawdowns

Reserve Component Drawdowns

The consequences of DoD personnel reductions at the end of the Cold War and in recent years were not as severe for reserve forces as they were for the active components. For example, the Army National Guard shrank by about 20 percent and the Army Reserves by about 33 percent from their peaks in 1991 to 1999. In comparison, the Army's active force reduced by about 39 percent from its peak in 1987 to 1999. The cutbacks to the reserve component that occurred in the 1990s were significantly larger than those that have taken place more recently although, in both cases, reductions have hit some service reserve components (particularly Army) harder than others.

The services' strategies for reducing reserve personnel appear to have varied somewhat across components and drawdown eras. Restricting accessions appears to be the most common method of meeting the so-far limited reduction goals of the current drawdown period. Force structure changes constitute key elements of the force-shaping strategies of the two National Guard components and the Navy Reserve, but not of the Army and Air Force Reserves.

The limited demographic data we were able to analyze suggest that female and minority representation generally increased across all of the reserve components during the 1990s. We did not receive demographic data for recent reductions, except from the Army National Guard. Based on additional data from the Army National Guard for FY 2010 to July 2014, female and minority representation increased anywhere from 0.4 to 1.9 percentage points. However, during the same period, enlisted blacks and women have been disproportionately subject to administrative (involuntary) separations. Although it is not clear what this trend in the Guard augurs, the extent to which reserve force reductions are focused on support units and new accessions (which tend to be more diverse than the overall military population) could determine the impact that downsizing efforts have on demographic diversity within the reserves.

Defense Civilian Drawdowns

The 1990s civilian drawdown was larger than planners initially expected. Between FY 1989 and 1999, the number of full-time civilian positions declined by about 400,000, from approximately 1,117,000 to 714,000—a 36-percent reduction (Brostek and Holman, 2000). Cuts were about equally distributed across the services. Although the size of the current civilian drawdown is likely to be smaller than the previous one, the Army could experience substantial reductions in the coming years, while the Air Force and Navy are expected to face more limited cutbacks.

The services' strategies for reducing their civilian workforces remain basically the same today as in the 1990s. Specifically, the services implement hiring freezes and offer voluntary incentives and early retirements initially. Then they institute involuntary reductions in force sparingly and only when necessary. They wish to avoid furloughs if at all possible.

While neither drawdown appears to have significantly changed the overall composition of the civilian workforce, members of certain groups seem to have been disproportionately affected. In the 1990s, these groups included blue-collar and clerical workers, junior employees, and, possibly, minorities and women. Recent involuntary separations have somewhat disproportionately impacted black men and women in the Army. For example, black civilians made up 17.2 percent of the Army civilian workforce between FY 2011 and the second quarter of 2014, but they represented 24.9 percent of civilians involuntarily separated in that period. For its part, the Air Force is concerned about a recent rise in mostly voluntary departures of Hispanic, female, and disabled civilians. The Navy does not expect its limited number of civilian reductions to have a disproportionate impact on any demographic group. One factor that could magnify the differential demographic effects of the current drawdown is the federal government's preference for hiring veterans, the majority of whom are white men, which could lead to a decreasing percentage of women and certain minorities in the ranks of DoD's civilian workforce.

Organization of This Report

The remaining chapters describe the findings, conclusions, and recommendations from our analysis of the impact of force reductions on demographic diversity. Chapter Two provides an overview of the strategies used by the services to reduce the sizes of their active military forces during drawdowns in the 1990s and the demographic diversity trends during that decade. Chapter Three is similar to Chapter Two but focuses on the Navy and Air Force drawdowns of the mid-2000s. Chapter Four transitions the report toward the present by outlining law, policy, and plans for recent force reductions. This chapter lays the groundwork for Chapter Five, which describes the methodology, results, and policy implications of our analysis of the potential impact of drawdown strategies on demographic diversity. Chapter Six summarizes the study findings and provides recommendations based on those findings.

CHAPTER TWO

Active-Duty Drawdown in the 1990s

In this chapter, we describe findings from our review of this drawdown period with a focus on demographic diversity. After we describe our approach to the review, we provide an overview of the goals and strategies that the services used to reduce their active-duty forces during the drawdown and the legal fallout from attempts that the Army and Air Force made to maintain demographic diversity during the drawdown. We then describe force structure changes from the drawdown. We follow this description with a discussion of our analysis of demographic diversity changes during and after the drawdown, using gender diversity in the Army officer corps as an example.

Approach

To identify information about the services' drawdown goals and strategies during the 1990s drawdowns, we relied on a combination of published literature and news reports, interviews with subject matter experts, and analyses of archival personnel data. To examine demographic trends during the 1990s, we primarily relied on our analyses of archival personnel data. Below, we provide a brief overview of our information sources. Additional details on our approach are in Appendix C.

Review of Published Sources on Drawdowns
We began our document search by focusing on literature about military drawdowns and demographic diversity. We searched Google Scholar

for publications between June 1989 and January 2014. We combined terms related to a military drawdown (e.g., "military downsizing," "military drawdown," and "reductions in force") with terms related to diversity (e.g., "diversity," "equal opportunity") and/or demographic categories (e.g., "women" and "minority"). Once we identified relevant sources through online search, we reviewed their reference lists to identify additional sources (i.e., snowball method). We also asked colleagues and interviewees if they could recommend other reports, articles, or documents relevant to the study. We found fewer than 20 sources on diversity and drawdown.

To fill in gaps on drawdown laws, policies, programs, and outcomes, we cast a wider net by searching additional databases other than Google Scholar and by not using search terms specifically tied to demographic diversity. We also searched for fact-based news articles (i.e., not opinion pieces) that addressed DoD and service-level drawdown policies, strategies, programs, and outcomes. We supplemented our electronic searches with snowball methods; we read reference lists of reports to identify other sources and asked colleagues and experts we interviewed for recommendations on publications to include.

Interviews

From fall 2013 through summer 2014, we conducted semistructured interviews with 55 subject matter experts to learn about past and recent active-duty military drawdowns.[1] Most of the experts have leadership roles in manpower and personnel policy organizations in OSD and the services (e.g., service "one shops" such as Army G-1 and offices in Manpower and Reserve Affairs, or M&RA). Most of the interviewees are officers at O-5 rank or higher or civilian personnel at a GS-15 grade-equivalent or higher.

We asked interviewees about their current positions and their roles (if any) during past drawdowns. If an interviewee could speak to both past and current drawdowns, we asked a series of questions about

---

[1]   We held similar discussions with experts on reserve component and defense civilian reductions. See Appendix A for the reserve component drawdown and Appendix B for the civilian drawdown.

the goals, strategies, and policies for the different drawdown periods. Otherwise, we asked about the drawdown period(s) for which they were knowledgeable.

Archival Personnel Data Analysis

For most of our analyses, we used historical personnel data from DMDC's Active-Duty Personnel Master File. This file includes annual snapshots of personnel records. Our analysis used information on the following personnel categories: demographics (e.g., gender, race/ethnicity, education level), entry information (e.g., date of entry, Armed Forces Qualification Test [AFQT] for enlisted personnel, accession source for officers), service background (component, corps [enlisted, warrant officer, commissioned officer]), career background (occupation, rank, years of service [YOS], time in grade [TIG]), and separation. We identified individuals who separated as those who were no longer in the data set a year after having a record in the data set.

In addition to DMDC data, we used summary data from the *Population Representation of the Military Services* ("PopRep"; DoD, 2011). The Office of the Assistant Secretary of Defense for Personnel and Readiness publishes PopRep results each year. We used the DMDC and PopRep data to describe annual demographic trends of the active military force by FY, service, and corps in the 1990s. We decomposed the overall annual demographic changes into changes in inflows (accessions) and outflows (separations) of personnel. The equation for the decomposition is described in Appendix C.

We supplemented the decomposition of the population changes by examining differences in retention rates between gender groups and racial/ethnic groups (i.e., non-Hispanic white versus non-Hispanic black and non-Hispanic white vs. Hispanic). First, we calculated unadjusted continuation rates by YOS, service, corps (commissioned officers and enlisted), and period (during drawdown [from FY 1990 to 1998] and after drawdown [from FY 1999 to 2001]). We present the continuation rates as cumulative continuation rates (CCRs). CCRs represent the probability that an active (officer/enlisted) accession will stay on active duty in a given service throughout the given year of service.[2] To compare CCRs across the two periods, we created synthetic cohorts by combining the accessions data from several cohorts in each period. Next, we used logistic regression models to adjust the CCRs for workforce characteristics that may differ between demographic groups: gender, race/ethnicity, AFQT (enlisted only), accession source (officers only), education, occupational category (nontactical operations vs. tactical operations), rank category[3], TIG, and period (FY). The adjusted CCRs allow us to conclude with more confidence that demographic differences in continuation trends are due to demographic group membership, not workforce characteristics.[4] However, we note that our adjustments do not distinguish between voluntary leavers, induced leavers, and involuntary separations. Appendix C provides more details about our data and approach.

---

[2]   Enlisted retention trends are more commonly represented by reenlistment rates for different terms of service. Because of limitations in our data prior to FY 1994, we could not compute these rates.

[3]   We grouped ranks into conceptually meaningful categories to reduce the number of parameters that the model would estimate. For enlisted personnel, we created three categories: junior (E-3 to E-4), mid-level (E-5 to E-6), and senior (E-7 to E-9). For officers, we created two categories: company grade (O-3 to O-4) and field grade (O-5 to O-6). For both officer and enlisted, we dropped the lowest ranks because most of the personnel are in training, so most separation reasons would not apply to them. For officers, we also excluded general/flag officer ranks (O-7 to O-10) because of their small numbers, particularly for women and racial/ethnic minorities.

[4]   Unlike the annualized decomposition changes, our CCR analysis involves the combination of multiple years of data to compare retention rates during a drawdown period to retention rates during the immediate postdrawdown period. We did this type of combination for two reasons. First, we did not have substantive and policy questions about year-to-year variations in retention behavior. Our main policy question is whether retention behavior under one policy regime (drawdown years) compares to retention behavior when that regime is no longer in place (postdrawdown years). Second, using discrete-survival models to capture year-to-year variations in retention rates across demographic groups can result in complex regression models that are difficult to interpret.

Goals and Strategies for the 1990s Drawdown

As in any postwar drawdown, the post–Cold War drawdown was driven by Congress' goal to cut military spending. DoD strove to reduce the total active force levels by 25 percent between FY 1987 and 1997 (Schroetel, 1993). Beyond the budgetary goal, the services had other goals. For example, the services wanted to ensure a trained and ready force remained after the drawdown. The Army, in particular, wished to avoid a "hollow force"—with its low-quality recruits, poor retention, and undermanned units—akin to the one the Army experienced after Vietnam (McCormick, 1998). Other Army goals included balancing officer and enlisted skills, maintaining enough accessions to sustain the force, controlling senior enlisted-grade growth, and ensuring adequate transition assistance to departing service members (McCormick, 1998). The services did not have explicit goals tied to demographic diversity, although McCormick notes that Army leaders had concerns about women and racial/ethnic minorities leaving the service. Given the use of equal opportunity language in some Air Force retention and promotion board instructions, Air Force leaders at the time shared the Army leadership's concerns.

Although maintaining demographic diversity was not an explicit drawdown goal, the one goal that the services, OSD, and Congress shared was "keeping faith" with existing personnel by not cutting their numbers dramatically, particularly through the use of involuntary separation measures. In the FY 1991 National Defense Authorization Act (NDAA), Congress specified drawdown priorities promoting the "keep faith" goal (Schroetel, 1993):

1. Reduce new (non–prior service) accessions first.
2. Reduce the portion of the force with more than 20 years of service by increasing retirements.
3. Limit the numbers of personnel with between two and six years, that is, limit the numbers entering the career force.
4. Involuntary separations are a last resort to be used only after the preceding measures have been taken.

Accession Cuts

The services largely followed the guidance that Congress laid out in FY 1991. The primary strategy was to cut accessions. As seen in Figure 2.1 (active-duty enlisted) and Figure 2.2 (active-duty officer), the services cut accessions dramatically between FY 1989 and 1991. The Air Force had the largest drops during that period (about 31 percent for enlisted and 32 percent for officer). Enlisted accession reductions continued for all but the Marine Corps through FY 1995. Officer accession cuts stopped around FY 1993, with some dips later in the decade, except for the Marine Corps.

Reducing accessions was economical and kept faith with existing service members. However, the large accession cuts contributed to an increase in the seniority of the services. On the enlisted side, the Air Force and Navy became more senior. Specifically, the proportion of personnel with ten or more YOS increased for most of the 1990s in those two services, dipping somewhat in the last part of the decade. In contrast, the proportion of enlisted personnel with ten or more YOS

Figure 2.1
Non–Prior Service Active-Duty Enlisted Accessions, by Service (FY 1989–2001)

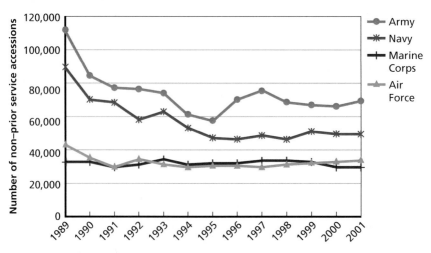

SOURCE: Data from Table D-4, DoD (2011).
RAND RR1008-2.1

Figure 2.2
Active-Duty Commissioned Officer Accessions, by Service (FY 1989–2001)

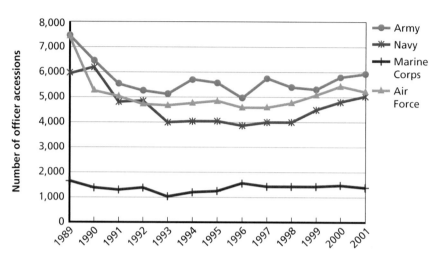

SOURCE: Data from Table D-15, DoD (2011).
RAND RR1008-2.2

began to decrease in the Marine Corps after 1993 and in the Army after 1994 (Rostker, 2013). On the officer side, from FY 1987 to 1997, all services gained officers in the 15 or higher YOS group, and all services except the Marine Corps lost ground in the eight or less YOS group. The Navy officer corps became most senior, increasing mid-level officers (eight to 14 YOS) 5 percentage points and senior-level officers (15 or more YOS) by 4 percentage points (Congressional Budget Office [CBO], 1999). As YOS increased, so did pay grade. Overall, the percentage of officers in the O-4–O-6 grades rose over 3 percentage points from 1989 to 1996, but the percentage of officers in the O-1–O-3 grades fell by 3.3 percentage points during the same period (CBO, 1999). To retain the more senior officer corps, Congress passed Defense Officer Personnel Management Act (DOPMA) relief on field-grade officer ceilings in 1995 and 1996. All services but the Army used the DOPMA relief to retain more field-grade officers.

The deep accession cuts that led to a more senior force may have created some challenges for some of the services. Senior forces are more expensive to maintain and can result in promotion stagnation for more junior personnel. Heavy accession cuts also create manpower "bathtubs" years later when the small entering cohorts reach the mid-level grades. Without enough mid-level personnel, the services have to rely on more-senior personnel to do the work that mid-level personnel would otherwise have done (Rostker, 2013). Senior personnel also carry an increased training burden when there are too many inexperienced personnel in their units. Indeed, Conley et al. (2006) found that the training loads increased in units for several Air Force enlisted specialties after the 1990s drawdown, resulting in senior personnel spending more time training others and less time doing their work. Another workload problem faced by the Air Force after the 1990s drawdown involved the high number of deployments. Specifically, several combat support units did not have enough personnel to cover the workload at the installations after others were sent on deployments (Conley et al., 2006).

Voluntary Incentives

In addition to accession cuts, the services used incentives to induce voluntary separation of existing personnel. Use of voluntary incentive packages also helped achieve the goal of "keeping faith" with service members by not forcing them to leave. In FY 1992, Congress funded the Voluntary Separation Incentive (VSI) and Special Separation Benefit (SSB) to target mid-career personnel who otherwise might have remained in the service until retirement (Mehay and Hogan, 1998). Because VSI and SSB were not popular among personnel nearing retirement eligibility, Congress authorized Temporary Early Retirement Authority (TERA) in FY 1993 (Rostker, 2013).

The services offered VSI and SSB between January 1992 and October 1995 (Asch and Warner, 2001). Service members could choose between the two packages: VSI offered an annuity for a set number of years, and SSB offered a lump-sum amount. The lump-sum package (SSB) was more popular than the annuity (VSI), with 90 percent of enlisted personnel and half of the officers choosing SSB over VSI (Rostker, 2013). However, the take-up rates for the two programs varied by service. The Army and Air Force had higher take-up rates than the Navy and Marine Corps. The Army and Air Force targeted the incentives at personnel who would otherwise be subject to involuntary reduction in force (RIF) boards (U.S. General Accounting Office [GAO], 1993). The Navy, in contrast, explicitly stated that RIFs would not occur if personnel did not take VSI or SSB (Mehay and Hogan, 1998; Asch and Warner, 2001). The young nature of the Marine Corps force allowed the Marine Corps to do much of its drawdown through accession cuts and restrictions on reenlistment past the first term.

Unlike VSI and SSB, TERA targeted service members with between 15 and 20 years of service for early retirement (Henning, 2006, and Wyatt, 1999). TERA provided an opportunity for service members to retire early and receive a reduced annual retirement package based upon the amount of time left until they reached 20 years of service. In total, 7,554 officers opted to accept TERA and separate from the services between 1991 and 1996. This amounts to approximately 0.5 percent of all officers who separated from service between 1991 and 1996 (CBO, 1999).

Econometric analyses of the effects of VSI and SSB on separations during the 1990s drawdown identified demographic group differences.[5] Racial/ethnic minorities eligible for VSI or SSB were less likely to accept an incentive package than non-Hispanic white personnel, whereas eligible women were more likely than eligible men to accept a package (Asch and Warner, 2001; Kirby, 1993; Miller, 1995). For the study on Army enlisted personnel, fewer women than men were eligible for one of the packages but more non-Hispanic black per-

---

[5]   Our review of economic factors is limited to the few studies that cover demographic trends in uptake of voluntary separation incentives. Other econometric studies examine the role of pay and bonuses during a drawdown period but do not cover demographic trends. For example, Hansen and Wenger (2002) found that Navy sailors were more responsive to pay when making reenlistment decisions during the 1990s drawdown than sailors before the drawdown and after the drawdown. The authors note that it is difficult to estimate the impact of reaction to pay during a drawdown period if there is not a good way to control for the change in service members' expectations during that period. According to Hansen and Nataraj (2011), an individual's decision to leave service after his/her service obligation is over is a function of factors including propensity for service, civilian job opportunities, relative compensation (i.e., military versus civilian), quality of life, family considerations, and job characteristics (e.g., work hours). Ultimately, an individual's decision to leave is based on that person's expectations regarding compensation and benefits, not necessarily the realities of compensation and benefits.

sonnel than non-Hispanic white personnel were eligible for the package (Asch and Warner, 2001). These demographic effects likely had limited effects on the overall demographic profile of the force during the drawdown, since VSI and SSB had a modest effect in inducing separations during the drawdown (Mehay and Hogan, 1998). Most of the reduction came from accession cuts.

Besides financial incentives, the services used early-out programs to induce personnel separations. For example, Congress reduced TIG requirements for officers with 20 or more YOS to induce voluntary retirements (GAO, 1995). To induce voluntary separations among enlisted personnel, the services used early release programs for personnel with less than six YOS, and in some cases, for personnel in overage specialties. Early release programs allowed personnel to leave before the end of their contracted term of service, usually within the same year. Because these programs did not come with severance pay, their popularity waned once VSI and SSB became available in 1992 (GAO, 1995).

Involuntary Measures

Although used to a lesser extent than voluntary measures and accession cuts, involuntary measures were employed by the services to reduce their forces in the 1990s. A variety of tools were used, the two most notable being Selective Early Retirement Boards (SERBs) and RIF boards.

Selective Early Retirement is a board process that targeted officers under certain criteria for early retirement. As prescribed by DOPMA, service secretaries have the authority to convene SERBs whenever necessary, and each service has its own service-unique policy to give guidance on implementing selective early retirement. Although SERBs could target officers in grades O-5 and up, the services primarily targeted O-5s and O-6s who met the criteria (i.e., twice passed over for promotion for O-5s and at least four years' TIG for O-6s).[6] The authority limits review of the same officer to every five years, and service secretaries could not recommend more than 30 percent of any grade for

---

[6]   Current law outlining SERB requirements resides in Title 10 of United States Code, Section 638.

separation by a SERB. The services limited their use of SERBs to avoid breaking faith with career officers (CBO, 1999).

Like SERBs, RIFs use a board process to select officers for involuntary separation. In the FY 1991 NDAA, Congress authorized the use of RIFs for officers with regular commissions in the O-3 and O-4 grades (CBO, 1999). The services already had authority to use RIFs for officers with reserve commissions.

In addition to SERBs and RIFs, the services limited promotion opportunities for officers. The services tightened the "up-or-out" provisions in DOPMA to force more officers to leave. Officers in the O-2, O-3, and O-4 grades twice passed over for promotion are forced to leave service. To be retirement eligible, officers had to achieve O-4 grade and 20 years of service. Tightening up-or-out provisions involved limiting which officers could be retained after being passed over for promotion. Although SERBs are a formal mechanism for executing up-or-out provisions, they are limited to officers in specific grades. The services did not rely solely on SERBs for enforcing up-or-out provisions.

For enlisted personnel, the services tightened reenlistment and tenure requirements to increase separations. First-term personnel wishing to reenlist faced stricter standards, such as those for physical fitness and weight (GAO, 1995). The services also used High Year Tenure (HYT) to limit who could stay in the service. HYT limits the years of service that an enlisted person can have in a particular pay grade without being promotion eligible (Wyatt, 1999). As with officers facing DOPMA career timeline restrictions, enlisted personnel facing HYT limits needed to move up (i.e., promote) or out (i.e., separate or retire). The HYT rules manifest as a reduction of retention control points, which establish the maximum number of years an enlisted member at a certain rank can stay before being denied the chance to reenlist (GAO, 1995). For example, the length of time that an E-4 could remain in service went from 13 years to ten years at the beginning of the drawdown (Asch and Warner, 2001). Tightening tenure rules for personnel who were retirement eligible signaled the possibility of forced retirement, which may have encouraged retirement-eligible personnel who were not likely to promote to elect an early separation incentive (Mehay and Hogan, 1998).

Diversity Goals and Legal Challenges

Based on our interviews with experts about the 1990s drawdown, the services had few explicit goals to ensure demographic diversity did not decrease due to the drawdown. The only explicit examples of diversity goals embedded in force-shaping decisions are the use of equal opportunity language in SERB, RIF, and promotion board instructions in the Army and Air Force. Army leaders at the time, such as former Undersecretary of Army John Shannon, were concerned that separation processes would adversely affect the demographic balance within the service. To address this concern, senior Army leaders added equal opportunity language to instructions for SERBs, RIF boards, and promotion boards. The language asked board members to be mindful that women and minorities may have experienced barriers in their careers that limited their opportunities for key assignments (e.g., command positions) that would make them as competitive as their white male peers. The Air Force senior leadership included similar language in SERB and RIF board precepts in the early 1990s.

Both the Army and Air Force experienced legal fallout from their attempts to limit negative demographic impacts of separation and promotion board processes. We briefly describe four of the lawsuits below (MLDC, 2010b).

- *Baker v. United States* (1995). A group of white Air Force colonels filed a reverse-discrimination lawsuit based on the results of a 1992 SERB. Col Baker and the other plaintiffs argued that the instructions to the SERB created a preference for women and minorities. The case largely focused on how the board interpreted the instructions but was ultimately settled.
- *Christian v. United States* (2000). This case also involved a SERB, but for lieutenant colonels in the Army. Like the *Baker* plaintiffs, plaintiffs in the *Christian* case argued that the instruction to the SERB created preferential treatment for women and minorities. The court ruled for the plaintiffs, thus not accepting the Army's two-pronged compelling government interest argument (1) "to create perceptions of equal opportunity in the Army" and (2) "to prevent possible past discrimination from negatively affecting

the present consideration of the officers' professional attributes and potential for future contributions if retained on active duty" (MLDC, 2010b, p. 3). The court did not accept the first part of the argument as valid (too amorphous) and thought the Army provided insufficient evidence of past discrimination to justify a race-based classification policy.

- *Alvin v. United States* (2001). White Air Force colonels challenged a 1994 SERB. The court ruled that the instructions to the SERB were not neutral, making it more favorable for women and minorities. However, the plaintiffs requested that the court not only subject the board instruction to strict scrutiny but that it rule that the Secretary of the Air Force should not have issued such guidance because it was not narrowly tailored. The court decided not to rule that the Secretary could have used other, more narrowly tailored means. As such, race- or gender-based instructions to SERBs could still pass strict scrutiny.
- *Berkley v. United States* (2002). A group of Air Force officers released by a FY 1993 RIF board challenged the instructions provided to the board. The Court of Federal Claims denied the plaintiff's claim that the RIF board instruction used race or gender classification and therefore did not apply the strict scrutiny test. However, the Court of Appeals for the Federal Circuit (2002) overturned the earlier court's decision and remanded the case back to the Court of Federal Claims (2004) to undergo strict scrutiny. The court did not decide if the Air Force policy would have met the strict scrutiny standard.

These four cases have important lessons about the use of demographic diversity considerations in drawdown decisions (MLDC, 2010b). One lesson is that board instructions with considerations regarding demographic diversity are likely to undergo strict legal scrutiny if challenged. Another lesson is that board instructions could theoretically survive strict scrutiny, though none was upheld in these cases. A third lesson is that the courts require strong evidence of past discrimination or risk of present discrimination if the policy or program is not enacted. For example, the court in the *Christian* case did not consider evidence that promotion rates are lower for minority groups sufficient proof of discrimination.

Another important consideration for these cases is that none involved an argument of diversity as a strategic imperative for the military. The strategic imperative argument was successfully applied in the Supreme Court case of *Grutter v. Bollinger* (2003), which upheld the University of Michigan Law School's affirmative action admission policy in the interest of promoting diversity. Justice Sandra Day O'Connor used language from an amicus brief filed by retired Lt. Gen Julius Becton, Jr., et al. to argue that a diverse and qualified officer corps was needed to fulfill the military's primary mission of providing national security (MLDC, 2010a, p. 3). Although more recent Supreme Court decisions have ruled against the use of affirmative action in public higher education admissions (*Schuette v. Coalition to Defend Affirmative Action*, 2014, and *Fisher v. University of Texas*, 2013), the Supreme Court did not overturn the *Grutter* decision. Indeed, "forward-looking" arguments such as diversity as a strategic imperative have limited legal testing to date; most cases revolve upon remediating past discrimination by the government entity in question (MLDC, 2010a).

Force Structure After the 1990s Drawdown

In addition to becoming more senior, the active military force became more heavily focused on officers. Across the services, the ratio of enlisted to officer personnel went from 6.4 to 1 in 1989 to 5.7 to 1 in 1996 (CBO, 1999). The Army experienced the largest reduction in its enlisted-to-officer ratio, going from 6.2 to 1 in 1989 to 5.0 to 1 in 1996 (19 percent reduction). The Navy's ratio reduced by 14 percent, and the Air Force's ratio reduced by 11 percent in that time period. Only the Marine Corps retained the same ratio in 1996 as it had in 1989 (8.8 to 1).

In its 1999 review of the aftermath of the officer drawdown in the 1990s, the CBO's findings about the reduction in enlisted-to-officer ratios was met with arguments by the services that the larger enlisted cuts were driven by changing requirements. The services argued, for example, that higher joint-billet requirements required retaining more officers, and new weapons systems reduced the need for enlisted man-power. However, the CBO speculated that another reason that the ser-vices separated relatively more enlisted personnel than officers during the drawdown was because it was easier to do.

Not only was the force more officer-heavy, but the occupational mix of the officer corps shifted away from tactical operations. Using personnel data from the DMDC, we estimate that the share of tactical operations officers decreased by 4 percentage points (42 to 38 percent) between FY 1989 and 1999 while the share of officers in nontactical operations occupations increased from 58 to 62 percent in the same time period[7]. The enlisted force became more nontactical as well in the Army and Marine Corps, but not in the Air Force and Navy. At least in the case of the Army, reductions to tactical operations personnel were a function of the nature of force structure cuts during the drawdown. Specifically, the Army's active-duty infrastructure force (Table of Dis-tribution and Allowances, or TDA) shrank by 31 percent between FY 1987 and 1999, whereas the Army's active-duty operational force (Table of Organization and Equipment, or TOE) shrank by 43 percent (Brinkerhoff, 2000). McCormick (1998) cites different arguments for why the Army cut more heavily from operational units than infrastruc-ture units. One argument is that a minimum level of TDA structure (hospitals, training areas, educational facilities) was necessary to sup-port the active Army, regardless of its size.

---

[7] The DoD occupational codes include a category for "tactical operations officers," which include career fields in infantry, armor, artillery, aviation, surface warfare, and submarines. For this comparison, the other occupational categories were placed under the banner of "nontactical operations occupations." Nontactical operations occupations include career field categories such as health care, intelligence, scientists, and supply.

Demographic Diversity During and After the 1990s Drawdown

Despite a significant reduction in the size of the active military force in the 1990s, representation of women and racial/ethnic minorities increased during the decade. As Figure 2.3 shows, female representation increased from just under 11 percent to just under 15 percent. The Navy had the largest percentage-point increase (more than 5), and the Marine Corps had the smallest increase (1.4 percentage points) with a drop during most of the drawdown period (from FY 1991 to 1995). Although not shown in the chart, female representation increased by over 4 percentage points (10.8 to 14.9 percent) in the enlisted force,

Figure 2.3
Female Representation in the Active-Duty Military, FY 1990–2001

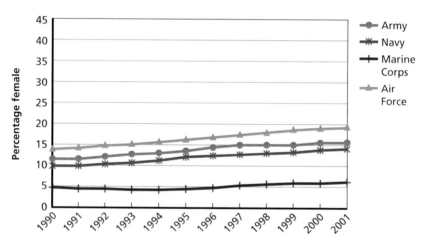

SOURCE: Analysis of enlisted and commissioned officer data from DMDC, FY 1990–2001.

RAND RR1008-2.3

and by more than 3 percentage points (12.1 to 15.3 percent) in the officer corps.[8]

Figure 2.4 shows racial/ethnic minority representation by service. Overall, minority representation increased by over 8 percentage points between FY 1990 (about 27 percent) and FY 2001 (about 35 percent). As with female representation, minority representation increased the most in the Navy (about 14 percentage points). Minority representation increased by nearly 9 percentage points (29.7 to 38.3 percent) in the enlisted force, and nearly 7 percentage points (10.4 to 17.1 percent) in the officer corps.[9]

Figure 2.4
Racial/Ethnic Minority Representation in the Active-Duty Military, FY 1990–2001

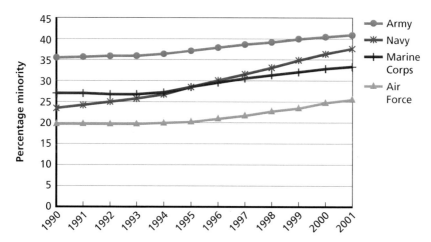

SOURCE: Analysis of enlisted and commissioned officer data from DMDC, FY 1990–2001.

RAND RR1008-2.4

---

[8] Female percentage-point gains by service: Army (4.5 for enlisted, 2.6 for officers), Navy (4.4 for enlisted, 4.0 for officers), Marine Corps (1.4 for enlisted, 2.1 for officers), and Air Force (5.6 for enlisted, 4.0 for officers).

[9] Minority percentage-point gains by service: Army (5.9 for enlisted, 5.7 for officers), Navy (15.3 for enlisted, 10.4 for officers), Marine Corps (6.4 for enlisted, 7.6 for officers), and Air Force (6.2 for enlisted, 4.7 for officers).

Although racial/ethnic diversity increased overall, non-Hispanic black representation decreased in the Army and Marine Corps by nearly 3 percentage points and over 4 percentage points, respectively. These reductions came from decreased black representation among enlisted personnel; black representation among Army officers was fairly stable and increased by almost 2 percentage points among Marine Corps officers. The losses in black representation among Army and Marine Corps enlisted were counteracted by gains for other minority groups, particularly Hispanics. Hispanics had gains of over 5 percentage points and 7 percentage points in the Army and Marine Corps enlisted ranks, respectively.

The overall increase in demographic diversity in the 1990s continued a trend from the 1970s (Quester and Gilroy, 2001). The inception of the All-Volunteer Force (AVF) is one factor credited with increasing the demographic diversity of the U.S. military since the 1970s (Rostker, 2006). Higher wages in the military relative to the civilian labor market and a merit-based promotion system also factored into increasing numbers of women and minorities in the military since the beginning of the AVF (Quester and Gilroy, 2001).

Example of Demographic Change in the 1990s: Gender Diversity in the Army Officer Corps

To understand what underlies the demographic trends of the 1990s active force, we unpacked Army officer trends to get an idea. We selected the Army because it is the largest of the services, and the recent Army reductions could result in the smallest Army in decades. We selected the Army officer corps because the Army's recent officer reductions—particularly in relation to their impact on diversity—have been in the media spotlight (see, for example, Vanden Brook, 2014).

We focus on gender diversity in the officer corps for ease of presentation (two gender groups versus multiple race/ethnicity groups). That said, we provide some analyses of demographic group differences based on the intersection of gender and race/ethnicity (e.g., non-Hispanic black women) to offer context for the main gender comparisons.[10] Additional information about other services, corps, gender, and race/ethnicity groups is provided in Appendix C.

Decomposition of Female Representation Changes

In Figure 2.5, we show the decomposition of annual change in female representation among active-duty Army officers. The figure shows total annual change in female representation (black bars) and the two components of total change: change due to gender mix of accessions (blue bars) and change due to gender mix of separations (light gray bars). Female representation among the officer corps increased in most years. The increases in female representation were driven by increased female shares of accessions and limited by female shares of separations, particularly in the mid-1990s (during the drawdown).

The female gains in accessions follow longer-term trends in female accession gains in the Army officer corps. Based on PopRep data on Army officer accessions, women represented about 11 percent of Army

Figure 2.5
Decomposition of Female Representation Changes, FY 1990–2001: Army Officers

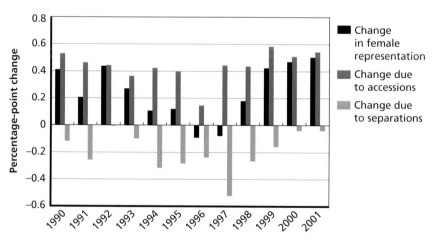

SOURCE: Analysis of DMDC data on active-duty Army personnel (FY 1990–2001).
RAND RR1008-2.5

[10] We considered additional analyses whereby we cross gender with race and ethnicity (e.g., black Hispanic women), but group sizes were too small for analysis in most cases.

officer accessions in FY 1974, 17 percent in FY 1980, and 18 percent in FY 1989. Except for a dip in FY 1996, female representation among Army officer accessions increased to nearly 22 percent by FY 2000 (Table D-16, DoD, 2011).

We also examined changes in representation for Army black officers and Hispanic officers. To save space, we produced Table 2.1 to show the decomposition trends for all of the gender-by-race/ethnicity groups (with negative trends bolded). In general, all demographic groups except for non-Hispanic white men enjoyed gains in representation for most of the years between FY 1990 and 2001. All of the female groups (white, black, and Hispanic) benefited from accession gains throughout the 1990s. However, non-Hispanic white women had consistently more separations than other groups, which is not the case for non-Hispanic black and Hispanic women. In terms of comparisons within race/ethnic groups, black female officers benefited relatively more from accession gains than from lower shares of separations, whereas black male officers would not have made overall gains by the end of the decade were it not for lower shares of separations compared to other groups. Increases in Hispanic officer representation were fairly balanced between increased shares of accessions and decreased shares of separations, though Hispanic female officers benefited more from accession gains than from lower separation losses.

Gender Retention Trends During and After the 1990s Drawdown

Because female representation in the Army officer corps during and after the 1990s drawdown was adversely affected by greater female shares of separations (particularly for white women), we examined these trends more closely by looking at gender differences in retention during and after the drawdown. Figure 2.6 shows the actual (observed) CCRs for male and female Army officers during and after the 1990s drawdown. The solid lines represent the drawdown period (from FY 1990 to 1998), and the dashed lines represent the postdrawdown period (from FY 1999 to 2001). Both during and after the main drawdown period, female retention is lower than male retention (orange line below blue line from same era). The gap is slightly larger during the drawdown years, with the largest drawdown-era gap occurring in years seven to

Figure 2.6
Army Officer Cumulative Continuation Rates, by Gender and Era,
FY 1990–2001

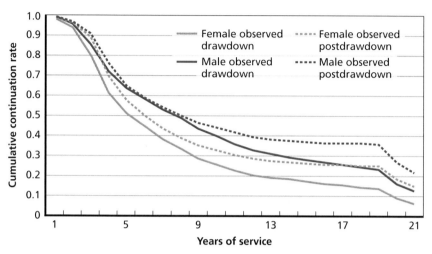

SOURCE: Analysis of DMDC data on active-duty Army officers (FY 1990–2001).
RAND RR1008-2.6

ten (14–15 percentage points) followed by years 11–17 (11 percentage points). The widest gap during the postdrawdown period is from seven to 18 YOS (11 percentage points).

The gender gap may be a function of differences between male and female officers that have more to do with force structure than gender. For example, the Army's female officers have been heavily concentrated in nontactical operations occupations, whereas the largest concentration of the Army's male officers has been in tactical operations occupations like infantry.[11] Table 2.2 shows that, other than the general officer corps, tactical operations occupations had the lowest representation of Army female officers in the 1990s (see gray shaded row). The 2.3 percent and 3.2 percent female representation in tactical operations occu-

---

[11] About 45 percent of the Army's male officers were in tactical operations occupations during the 1990s drawdown (from FY 1990 to 1998), and about 43 percent were in those occupations in the years immediately after the drawdown (from FY 1999 to 2001). The next largest concentration of the Army's male officers was in health care occupations (17 percent during drawdown and 18 percent after drawdown).

Table 2.1
Decomposition of Gender-by-Race/Ethnicity Group Representation Changes, FY 1990–2001: Army Officers

Demographic Group	1990	1991	1992	1993	1994	1995	1996	1997	1998	1999	2000	2001
NH White Men												
Accessions	-0.71	-0.57	-0.54	-0.27	-0.57	-0.50	-0.26	-0.53	-0.59	-0.67	-0.77	-0.84
Separations	-0.08	0.06	-0.22	0.06	0.17	0.13	0.16	0.44	-0.02	0.02	-0.18	-0.17
Total	-0.79	-0.51	-0.76	-0.21	-0.40	-0.37	-0.10	-0.09	-0.61	-0.65	-0.95	-01.01
NH White Women												
Accessions	0.33	0.30	0.31	0.29	0.28	0.26	0.05	0.30	0.26	0.33	0.27	0.23
Separations	-0.13	-0.26	-0.09	-0.07	-0.28	-0.24	-0.24	-0.44	-0.26	-0.19	-0.10	-0.16
Total	0.20	0.04	0.22	0.22	0.00	0.02	-0.19	-0.14	0.00	0.15	0.16	0.07
NH Black Men												
Accessions	0.01	-0.06	-0.10	-0.19	-0.08	-0.12	-0.09	-0.14	-0.09	-0.10	-0.05	0.02
Separations	0.13	0.14	0.12	-0.04	0.08	0.09	-0.01	0.00	0.15	0.05	0.13	0.10
Total	0.14	0.08	0.02	-0.23	0.00	-0.03	-0.10	-0.14	0.07	-0.05	0.08	0.13

Table 2.1—Continued

Demographic Group	1990	1991	1992	1993	1994	1995	1996	1997	1998	1999	2000	2001
NH Black Women												
Accessions	0.11	0.09	0.07	0.03	0.06	0.04	0.03	0.04	0.05	0.14	0.11	0.17
Separations	0.02	0.01	0.06	−0.04	−0.03	−0.03	0.01	−0.05	0.02	0.02	0.08	0.10
Total	0.13	0.10	0.14	−0.01	0.03	0.01	0.04	−0.02	0.07	0.16	0.19	0.27
Hispanic Men												
Accessions	0.06	0.03	0.08	0.02	0.09	0.07	0.08	0.05	0.08	0.10	0.14	0.07
Separations	0.10	0.10	0.16	0.07	0.07	0.07	0.04	0.03	0.07	0.04	0.07	0.10
Total	0.16	0.13	0.24	0.09	0.15	0.14	0.13	0.08	0.14	0.14	0.20	0.17
Hispanic Women												
Accessions	0.04	0.03	0.03	0.01	0.02	0.04	0.02	0.03	0.05	0.02	0.06	0.05
Separations	0.01	0.00	0.02	0.03	0.00	0.00	0.01	−0.02	0.01	0.01	−0.01	0.01
Total	0.04	0.04	0.05	0.04	0.02	0.04	0.03	0.02	0.05	0.03	0.05	0.07

NOTES: Each value in the table reflects a percentage-point change from the previous FY. Negative values are bolded to facilitate identification of trends. NH stands for "non-Hispanic."

Table 2.2
Representation of Army Female Officers in Occupational Categories During and After the 1990s Drawdown

Occupational Category	During Drawdown, % (FY 1990–1998)	After Drawdown, % (FY 1999–2001)
Administrators	24.0	26.6
Engineering and maintenance officers	16.0	18.8
General officers and executives	1.1	3.0
Health care officers	31.4	30.9
Intelligence officers	15.6	17.5
Scientists and professionals	10.1	14.7
Supply, procurement and allied officers	17.6	18.3
Tactical Operations Officers	2.3	3.2
Total (baseline)	14.1	15.4

SOURCE: Analysis of DMDC data on active-duty Army officers (FY 1990–2001).

NOTE: Occupational categories are based on those used by DMDC to compare occupational groups across services.

pations reflects significant underrepresentation of Army female officers, given they were about 14 to 15 percent of the Army officer corps in the 1990s and early 2000s. Women's underrepresentation in tactical operations career fields is mostly a reflection of law and DoD policy that has restricted women's assignment to combat positions, including entire career fields (e.g., infantry). Until 2013, women were restricted from assignment to direct ground combat positions. These restrictions meant that women's opportunities in tactical operations occupations were limited, or in several cases, nonexistent.[12] In contrast to the findings for tactical operations occupations, Army female officers were particularly overrepresented in health care occupations and administrative occupations. Women's occupational preferences, as well as "needs of the service," likely factor into their overrepresentation in these occupational areas.

Based on these descriptive findings, we hypothesized that occupational differences may partly explain gender differences in retention. To explore whether occupation and other relevant workforce characteristics could explain the gap, we adjusted the female CCRs to "look like" male CCRs during their respective eras (e.g., female drawdown CCRs adjusted to male drawdown CCRs). If workforce factors explain the gender gap, we would expect to see the adjusted female CCR line close to the male unadjusted (observed) CCR line. Figure 2.7 provides the adjusted female CCRs and the unadjusted male and female CCRs from the 1990s drawdown era and the postdrawdown era (1999–2001). The adjusted female CCR line for the drawdown era (dashed orange) stays close to the observed female CCR line for the drawdown era (solid orange) for the first six YOS (i.e., junior officers). However, the adjusted female CCR line begins to diverge from the observed female CCR line around seven YOS until it meets with the observed male CCR line for the drawdown era (solid blue) around 18 YOS. The widest gap between the drawdown adjusted female CCRs and observed male CCRs occurs between three and eight YOS.

---

[12] Law and policy on women's assignment to combat roles has evolved since the late 1980s. In 1988, DoD formalized a "risk rule" that banned women from assignment to units or occupations at "risk of exposure to direct combat, hostile fire, or capture that was equal to or greater than that of combat units in the same theater of operations" (Miller et al., 2012, p. 2). In 1993, Congress opened positions on combat aircraft and naval ships to women but ordered DoD to notify Congress if other combat positions were to be opened to women. In 1994, then–Secretary of Defense Les Aspin rescinded the 1988 "risk rule" and set new policy on the assignment of women. The 1994 policy (also known as the "direct combat exclusion" policy) restricted women from assignment to units below brigade level that have the primary mission of direct ground combat (Miller et al.). In 2013, the Chairman of the Joint Chiefs of Staff, General Martin Dempsey, and then–Secretary of Defense Leon Panetta issued a memorandum rescinding the 1994 policy and directing all closed positions be opened to women (unless approved by the Secretary and Chairman to remain closed) by January 1, 2016.

Figure 2.7
Actual and Adjusted Cumulative Continuation Rates by Gender During and After the 1990s Drawdown: Army Officers

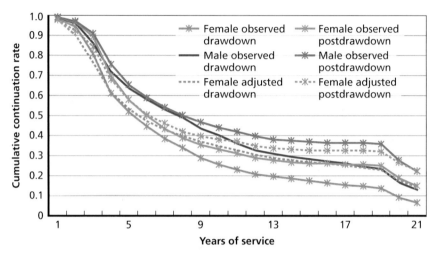

SOURCE: Analysis of DMDC data on active-duty Army officers (FY 1990–2001).
RAND RR1008-2.7

The pattern is similar for the postdrawdown years, but the gender gap is wider. The adjusted female CCR line for the postdrawdown era (dashed gray with stars) stays close to the observed female CCR line for the postdrawdown era (solid gray with stars) until around six YOS, after which the adjusted female CCR line gets closer to the observed male CCR line for the postdrawdown era (solid green with stars). However, the adjusted female CCR line does not merge with the male CCR line as it does in the drawdown years.

The trends in Figure 2.7 suggest that the gender difference in retention during and after the 1990s drawdown is a tale of two trends. One trend is for junior officers, for which the gender gap is not explained by differences between men and women in terms of rank, education, and other work-related characteristics, including occupational category. The other trend is for senior-level officers (i.e., officers at or near retirement). For these officers, most of the gender gap is due to workforce characteristics other than gender.

Although not tested in our models, we hypothesize that the gender gap for junior Army officers is about lack of fit with military life, which may occur for a host of reasons (family choices, geographic stability, etc.). However, for senior officers, the retention decision is tied closely to the up-or-out promotion system and the 20-year retirement system that puts "golden handcuffs" on personnel nearing retirement. The up-or-out system homogenizes officers as they increase in rank so that male and female officers become more alike in terms of workforce characteristics. The retirement system's "golden handcuffs" constrain retention behavior so that male and female officers who are nearing retirement eligibility behave much more similarly than junior female and male officers.

The gender gap in officer retention is not isolated to the 1990s Army officer corps. As we show in Appendix C, gender gaps in officer retention exist across services in the 1990s, though the gaps vary in size by service. We also find gender gaps in the Air Force and Navy in the 2000s drawdown era. In support of our findings of a persistent gender gap in officer retention, a recent RAND study found a gender gap in Air Force officer retention from FY 2001 to 2011 and even when controlling for workforce factors and personal life factors (e.g., marital status) (Lim et al., 2014). The gender gap in retention is therefore not limited to drawdown eras and persists to this day.

Race/Ethnicity Retention Trends for Army Female Officers

In addition to our main analysis of gender retention trends, we explored whether women of different racial/ethnic backgrounds differ in retention trends. We show in Table 2.1 that non-Hispanic white female officers had relatively more separations than other demographic groups, whereas non-Hispanic black female officers and Hispanic female officers did not show major separation losses relative to other groups. Figure 2.8 depicts the observed retention rates (CCRs) for non-Hispanic white, non-Hispanic black, and Hispanic women by drawdown period. These raw rates indeed show that non-Hispanic white women had lower retention rates than non-Hispanic black women and Hispanic women during and after the 1990s drawdown. In fact, the postdrawdown rates for white women (solid navy blue line) were lower

Figure 2.8
Army Female Officer Cumulative Continuation Rates, by Race/Ethnicity and Era, FY 1990–2001

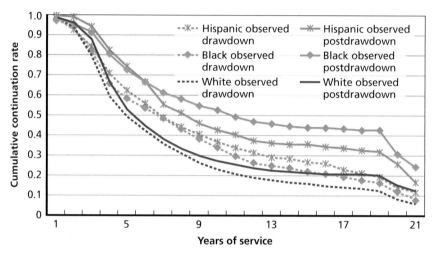

SOURCE: Analysis of DMDC data on active-duty Army female officers (FY 1990–2001).
RAND *RR1008-2.8*

than the drawdown rates for Hispanic women (dashed orange line with stars) and black women (dashed gray line with diamonds).

As with our main gender comparisons, we adjusted CCRs to determine if workforce characteristics such as education, rank, and occupation could explain the gap between the baseline group (in this case, white women) and the other groups (black women and Hispanic women). Figure 2.9 provides the adjusted black female CCRs and Hispanic female CCRs for the 1990s drawdown period, as well as the observed (actual) CCRs during the drawdown for white women (baseline), black women, and Hispanic women. Adjusting the post-drawdown black female officers (gray dashed line with diamonds) and Hispanic female officers (orange dashed line with stars) to "look like" white female officers during the drawdown (solid blue line) explains most of the gap among the three groups of women.

We also found that adjusting Hispanic women's and black women's CCRs for the postdrawdown era produced similar results to those shown in Figure 2.9, i.e., the adjusted CCRs are similar to the observed

Figure 2.9
Actual and Adjusted Cumulative Continuation Rates, by Race/Ethnicity, in 1990s Drawdown Era: Army Female Officers

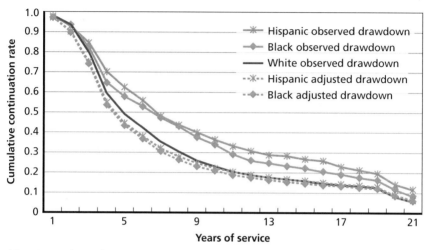

SOURCE: Analysis of DMDC data on active-duty Army female officers (FY 1990–2001).
RAND RR1008-2.9

CCRs for white women. Taken together, the results suggest that differences in occupation, rank, or other workforce structure characteristics explain most of the differences between women from different racial/ethnic groups.

Summary

Despite having to draw down its active-duty force after the Cold War, the U.S. military experienced an increase in demographic diversity from FY 1990 to 2001. This increase continued a trend beginning in the late 1970s, after the inception of the AVF. Nonetheless, policy decisions during the drawdown may have affected the demographic trends. The most notable policy decision of the drawdown was to slash accessions. The increasing female and minority shares of accessions allowed the female and minority populations to lose relatively fewer members than the male and white populations. We show such a trend for female representation in Army officer corps in the 1990s, with female accessions driving female representation increases. Lower female retention than male retention limited further growth in the female population. Except for senior-level Army officers (i.e., those near retirement), the gender retention gap during and after the drawdown is not fully explained by differences in other workforce characteristics, such as occupational category.

We also explored whether retention rates differed by race/ethnicity group for Army female officers. We find that non-Hispanic women had much lower retention rates both during and after the 1990s drawdown than non-Hispanic black women and Hispanic women in the Army officer corps. However, when the retention rates of black women and Hispanic women are adjusted to "look like" the retention rates of white women during the drawdown, we find that retention rates for the three groups of women are similar. Therefore, racial/ethnic differences in observed retention rates for Army female officers during and after the 1990s drawdown are mostly a function of group differences in rank, occupation, education, and other workforce characteristics.

Other drawdown policies besides accession cuts may have affected demographic diversity to some extent. However, teasing apart the relative influences of different drawdown policies on demographic diversity presents a challenge because the services used several drawdown policies and programs at the same time, targeted different segments of the population, and did not provide enough details on the sequence of events to allow a detailed analysis. The limited details on how drawdown strategies may have affected demographic diversity may also stem from a lack of diversity-oriented goals for the drawdown, with the exception of the Army and Air Force's use of equal opportunity language in officer separation board instructions. This strategy did not end well for either service; both services faced discrimination lawsuits from white male officers who were separated by such boards. As a result, the services have not since attempted similar strategies to maintain demographic diversity of the force.

CHAPTER THREE

Navy and Air Force Active-Duty Drawdowns in the Mid-2000s

In Chapter Two, we presented findings from the 1990s drawdown. Despite the services having few goals for demographic diversity during that drawdown, demographic diversity generally increased throughout the 1990s. In this chapter, we discuss the drawdown that occurred in the mid-2000s in the Navy and Air Force. Unlike the Army and Marine Corps, which grew in response to demands for ground troops in Afghanistan and Iraq,[1] the Navy and Air Force cut their forces. Most of the reductions occurred from FY 2003–2008 in the Navy and FY 2005–2008 in the Air Force, although both services continued force-shaping measures beyond FY 2008.

This chapter presents goals, strategies, and force structure changes from the drawdowns of the Navy and Air Force in the mid-2000s. We use the same data sources and analyses as in Chapter Two. Please refer to that chapter's methodology discussion and Appendix C for details on the approach used for this chapter.

Goals and Strategies for the Mid-2000s Drawdowns

The Navy and Air Force goals in the mid-2000s drawdowns did not dramatically differ from goals used for the 1990s reductions. The one exception is that the goal to "keep faith" with the career force was less

---

[1]   Between FYs 2002 and 2010, the active enlisted force grew by 13 percent in the Army and 14 percent in the Marine Corps. The Army active-duty officer ranks grew nearly 16 percent, and the Marine Corps active-duty officer ranks grew about 15 percent.

prominent in the mid-2000s reductions. Relative to the 1990s, both services relied less on accession cuts and more on separation strategies that would balance occupations (skill areas) and experience levels while retaining high-quality personnel.

Navy

The Navy began to reduce its active forces in FY 2003. In 2002, the Navy's active enlisted force was around 324,700. By FY 2011, it was down to 266,900, a drop of nearly 18 percent. The Navy's active officer corps also shrank in size, but over a shorter period of time (FY 2003–2008) and by a smaller amount (6.7 percent; from 53,323 in 2003 to 49,735 in 2008).[2] The Navy cited decommissioning of "manpower-intensive ships" and efficiencies in newer ships as reasons for the reductions (CBO, 2006, p. 53).

Based on our interviews with Navy drawdown experts, the Navy did not have explicit diversity goals for its drawdown. Instead, the Navy focused on maintaining a balanced and qualified force without directly considering the impact of its decisions on demographic diversity. To achieve its goals for a balanced force, the Navy relied more heavily on involuntary separation measures and less on accession cuts than in the 1990s. For example, the Navy cut enlisted accessions by about 49 percent between FY 1988 and 1995. In contrast, the Navy cut accessions by only 21 percent from FY 2002–2007. In terms of involuntary separation measures, the Navy used a combination of reenlistment controls and retention boards to reduce large numbers of enlisted personnel. In 2003, the Navy began the Perform-to-Serve program, which targeted mid-grade personnel in overage ratings and particular year groups who were nearing the end of their enlistment contracts. The Navy used performance-based criteria to determine which sailors could reenlist. However, sailors not selected to reenlist could apply to stay in the Navy by switching to undermanned ratings. Over 7,000 sailors left the Navy under this program in 2011 (Reilly, 2012).

Although the Navy relied on the Perform-to-Serve program to help meet its end-strength goals, overage ratings still maintained very

---

[2]   Enlisted and officer end-strength estimates based on Tables D-11 and D-17, DoD (2011).

high retention levels, especially from FY 2008–2010. As a result, the Navy used Enlisted Retention Boards (ERBs) to separate more sailors from overage specialties. However, this time, personnel in mid-contract (i.e., not near a reenlistment point) were targeted. The Navy held two ERBs in mid-to-late 2011, resulting in nearly 3,000 sailors cut from service (Faram, 2011).

Air Force

As with the Navy, DoD informed the Air Force in late 2003 that it needed to reduce the size of its force to achieve the required end-strength of 359,700 by the end of FY 2005. In FY 2004, the Air Force began force shaping and, by the end of FY 2005, exceeded its goal, reaching an end-strength of approximately 353,700. However, in December 2005, DoD released Program Budget Decision (PBD) 720 that mandated the Air Force to reduce its end-strength to about 316,000 by the end of FY 2011. The Air Force initiated a new round of force-shaping programs that ran through the rest of FY 2006 and into FY 2008. In early 2008, then–Secretary of Defense Robert Gates directed the Air Force to end further end-strength cuts. As of March 31, 2008, the Air Force had 328,600 personnel.

Unlike the Navy reduction, the Air Force reduction was more balanced between the officer and enlisted forces. Specifically, the active-duty enlisted force was at a seven-year high of 298,300 in FY 2004 and shrank down to 258,100 by FY 2008, a reduction of over 13 percent. The officer corps was also at a seven-year high in FY 2004, reaching 74,304. It lost nearly 9,500 officers by FY 2008, a reduction of about 13 percent.[3]

As in the 1990s, the Air Force achieved a portion of its drawdown through accession cuts. However, accession cuts in the 2000s drawdown were not as severe as in the Cold War drawdown period, with one exception. The Air Force sharply cut enlisted accessions from around 33,700 in FY 2004 to just over 19,000 in FY 2005. The Air Force corrected course in FY 2006 out of concern about an imbalance of officer-to-enlisted personnel; enlisted accessions returned to just

---

[3]   See DoD (2011).

over 30,000 in FY 2006, while officer accessions diminished by 600 (U.S. Air Force Audit Agency, 2008, p. 4). Air Force enlisted accessions in FY 2006 accessions were only 9.7 percent lower than they were in FY 2004. In contrast, between FY 1989 and 1991, the Air Force cut enlisted accessions by 31 percent.

To achieve the rest of its reductions, the Air Force used a variety of voluntary and involuntary measures. Based on estimates provided by an Air Force force-shaping policy expert, over 19,500 enlisted and officer personnel left under a voluntary program between FY 2004 and 2008. In comparison, over 9,500 enlisted and officer personnel left via involuntary separation programs during the same period. Voluntary measures included transfers from the Regular Air Force to the Air Force Reserve Component (ARC) via the Palace Chase program, transfers to the Army via the Blue to Green program, limited waivers of Active Duty Service Commitments, TIG waivers, Reserve Officers' Training Corps commissions to the ARC or Army (officers only), commissioned service waivers (officers only), and Voluntary Separation Pay (VSP; officers only). Involuntary separation programs included SERBs, RIFs, and Force Shaping Boards for officers, and reenlistment controls for enlisted personnel.

The Air Force had two reenlistment control programs, Date of Separation (DOS) Rollback and Career Job Reservation (CJR) Limited. DOS Rollback allowed commanders to accelerate the DOS of Airmen with specific reenlistment eligibility codes (e.g., Article 15). DOS Rollback was limited to personnel with 14 years or less total service at time of separation. CJRs determine the availability for first-term Airmen to reenlist in their current skill based on quality factors and career field positions available. Under normal circumstances, CJR approval is automatic. Under force shaping, a number of specialties (typically overmanned specialties) were constrained so that CJR was not automatically approved. To determine CJR approval, Airmen were rank ordered on a set of factors, such as their grade, projected grade, last three performance reviews, and date of rank.

Like the Navy, the Air Force targeted voluntary and involuntary programs toward less critical specialties and overage year groups where possible. Likewise, the Air Force used personnel quality factors, such as performance ratings and disciplinary actions, for some programs. Another important similarity between the two services' drawdowns is that neither included explicit demographic diversity goals for force-shaping decisions. Instead, the services focused on shaping the force according to skill area, year group, and, where possible, personnel quality.

Force Structure After the Mid-2000s Drawdown

Based on its larger cut to the enlisted force than to the officer corps, the Navy had an active-duty force weighted more toward officers than enlisted by FY 2008. Specifically, the enlisted-to-officer ratio dropped from about 6.1-to-1 in FY 2002 to 5.5-to-1 in FY 2008. The Air Force, in contrast, did not experience a drop in the enlisted-to-officer ratio between FY 2004 and FY 2008 (ratio of 4.0-to-1 in both years).[4] As previously noted, the Air Force leadership was concerned about the enlisted-to-officer ratio dropping after the deep enlisted accession cuts in FY 2005 and, as a result, increased enlisted accession numbers in following years.

Although the Navy lost more sailors relative to officers, the Navy did not become more experienced as it did after the 1990s drawdown. Based on DMDC data, the share of Navy officers with 15 or more YOS dropped about 0.5 percentage points between FY 2002 and 2009. The share of enlisted personnel with 15 or more YOS dropped about 4.3 percentage points between FY 2002 and 2011. The Air Force enlisted force also became more junior after the main Air Force drawdown period, with the share of enlisted personnel with 15 or more YOS reducing about 3.4 percentage points between FY 2004 and 2009. However, the Air Force officer corps became somewhat more experienced; the share of those with 15 or more YOS increased about 1.6 percentage points in the same period. The biggest drop for the Air Force officer corps was in the mid-year groups (eight to 14 YOS), about a 1.6 percentage-point reduction.

---

[4]   See DoD (2011).

The occupational mix in the Navy and Air Force changed after the main 2000s drawdown periods ended. The officer corps in both services leaned slightly toward tactical operations. Using DMDC data, we estimated that the share of tactical operations officers increased by about 1.3 percentage points between FY 2002 and 2009 for the Navy and about 3.1 percentage points between FY 2004 and 2009 for the Air Force. Meanwhile, the share of officers in nontactical operations occupations decreased about 1.4 percentage points in the Navy and about 1 percentage point in the Air Force. Unlike the officer force, the enlisted force became more nontactical; the share of personnel in the occupational category "infantry, gun crews, and seamanship special-ists" decreased by about 0.3 percentage points between FY 2002 and 2011 in the Navy but by over 7 percentage points between FY 2004 and 2009 in the Air Force.[5]

Demographic Diversity During and After the Mid-2000s Drawdown

Unlike gender diversity in the 1990s, gender diversity did not dramati-cally increase in the Navy and Air Force in the 2000s. As Figure 3.1 demonstrates, the Air Force enlisted force lost a very small amount of gender diversity (0.5 percentage-point reduction), whereas female rep-resentation increased by 1.4 percentage points in the Air Force officer corps. Female representation increased in both the Navy enlisted force (by 2.4 percentage points) and officer corps (by 1.2 percentage points).

Figure 3.2 shows trends in racial/ethnic minority representa-tion. Minority representation increased more in the Navy enlisted force (11.6 percentage-point increase) than in the Navy officer corps (5.3 percentage-point increase). Much of the Navy's enlisted increase came from increased Hispanic representation, as shown in Table C.6 in

---

[5]   The "infantry, gun crews, and seamanship specialists" occupational category in DMDC data does not include intelligence, which the Air Force classifies as part of its operations personnel. Air Force intelligence occupations experienced a small increase during the draw-down: The DMDC occupational category "communications and intelligence specialists" increased by about 0.5 percentage points between FY 2004 and 2008.

Figure 3.1
Female Representation in the Active-Duty Navy and Air Force, FY 2001–2011

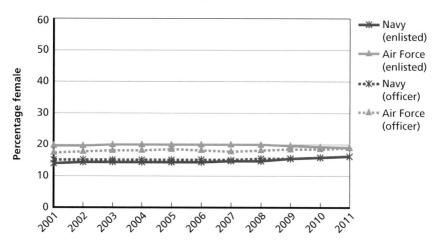

SOURCE: Analysis of enlisted and commissioned officer data from DMDC, FY 2001–2011.

RAND RR1008-3.1

Figure 3.2
Racial/Ethnic Minority Representation in the Active-Duty Navy and Air Force, FY 2001–2011

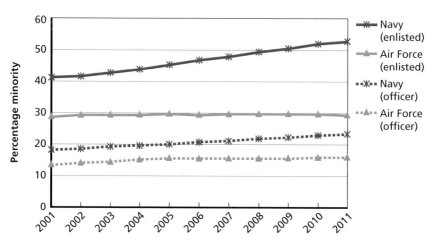

SOURCE: Analysis of enlisted and commissioned officer data from DMDC, FY 2001–2011.

RAND RR1008-3.2

Appendix C. In the Air Force, minority representation increased by 2.9 percentage points in the officer corps versus 0.9 percentage point in the enlisted force. Overall, minority representation increased by 10 percentage points in the Navy and 1.2 percentage points in the Air Force.

Example of Demographic Change in the 2000s: Gender Diversity in the Air Force Enlisted Force

As we did with the 1990s drawdown, we unpacked demographic trends from one service and one corps for the 2000s drawdown. We chose to focus on gender diversity in the Air Force enlisted population. We chose the Air Force instead of the Navy because the Air Force reductions had a more defined period because of the PBD 720 reduction from FY 2005–2008. We focus on the enlisted force instead of the officer corps because gender diversity decreased in the enlisted force while increasing in the Air Force officer corps and in the Navy. Appendix C provides tables with results for all services and corps by gender and race/ethnicity.

Decomposition of Female Representation Changes

Figure 3.3 shows the decomposition of annual change in female representation among active-duty Air Force enlisted personnel in the 2000s. In the years before the drawdown (pre-2004), female representation increased. However, for most of the drawdown years and the years that followed, female representation decreased as seen by the black bars below zero. For the entirety of the 2000s, women had a larger share of separations than men (light gray bars below zero). Women also had a larger share of accessions for most of the decade, as shown by the blue bars above zero. Female representation losses would have been larger had women had a lower share of accessions. The two years where female shares of accessions decreased, FY 2010 and 2011, had the largest drops in female representation.[6]

---

6   We also analyzed representation trends by race/ethnicity category overall and by race/ethnicity crossed with gender (e.g., non-Hispanic black female, Hispanic male). Although the magnitudes somewhat differ, the results for different race/ethnicity groups within gender category followed similar patterns. In general, Air Force enlisted black and Hispanic repre-

Figure 3.3
Decomposition of Female Representation Changes, FY 2001–2011: Air Force Enlisted

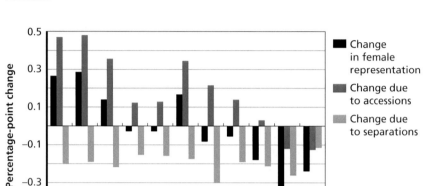

SOURCE: Analysis of DMDC data on active-duty Air Force enlisted personnel
(FY 2001–2011).
RAND RR1008-3.3

The question is, what happened to the female accession advantage? Based on our interviews with Air Force policy experts, at least two events at the beginning of the drawdown may have affected female accessions. One event is the sharp decrease in enlisted accessions in FY 2005. Accessions dropped from over 34,000 in FY 2004 to around 19,000 in FY 2005. The Air Force readjusted, and accessions were back up around 30,000 in FY 2006. The jump in female share of accessions in FY 2006 may be a function of the jump in accessions after the large dip in FY 2005.

The other event was the renorming of the Armed Services Vocational Aptitude Battery (ASVAB), a required test for entry into the military enlisted force. In 2004, DoD renormed the ASVAB. The Air Force tightened ASVAB enlistment standards after the renorming. Using DMDC data, we examined gender trends in AFQT scores, which are

---

sentation dropped mainly due to decreases in the shares of accessions, not increases in the shares of separations. For the full results, see Table C.10 in Appendix C.

composites of ASVAB[7] scores. Specifically, we examined AFQT scores for Air Force men and women with up to one YOS to approximate accession cohorts. We grouped the AFQT scores into the categories that the services use to make decisions on who can enter the service (Categories I, II, IIIA, IIIB, IV, and V).[8] In general, Air Force accession cohorts had increasingly higher AFQT scores throughout the 2000s, particularly after FY 2008. For example, the percentage of Air Force accessions with AFQT scores in the top two AFQT categories (I and II) increased from around 52 percent in FY 2008 to around 63 percent by FY 2011. Meanwhile, the proportion of Air Force enlisted accessions with scores in one of the lowest AFQT categories (IIIB) dropped precipitously after FY 2008, from around 20 percent to 5 percent by FY 2011. This drop may be due to the Air Force policy to enact stricter entry standards based, in part, on AFQT scores. Figure 3.4 shows the proportion of Air Force male and female accessions in AFQT Category IIIB by fiscal year. The female share of Category IIIB accessions is somewhat higher than the female share of Air Force enlisted personnel, as shown by the gray bars (female share of Category IIIB accessions) above the black line (female representation in force). To the degree that the Air Force restricted Category IIIB accessions, female representation could have decreased (all else being equal).

Gender Retention Trends During and After the 2000s Drawdown

Air Force enlisted women might not have lost representation had they retained at the same rate as males. As in Chapter Two, we examine actual (observed) and adjusted CCRs during and after the drawdown era for Air Force enlisted personnel. Figure 3.5 presents the observed CCRs for men and women during the main drawdown period (FY

---

[7]   The ASVAB has four subtests: Paragraph Comprehension (PC), Word Knowledge (WK), Mathematics Knowledge (MK), and Arithmetic Reasoning (AR).

[8]   AFQT scores are percentiles of the test-taking population and are grouped into rank-ordered categories, with Category I having the top AFQT scores. Category I and II recruits are highly sought-after by the services. AFQT categories have the following score bands: 93–100 (Category I), 65–92 (Category II), 50–64 (Category IIIA), 31–49 (Category IIIB), 10–30 (Category IV), 1–9 (Category V). The services are prohibited from accepting applicants in Category V.

Figure 3.4
Gender Representation of AFQT Category IIIB Accessions, FY 2002–2011:
Air Force Enlisted

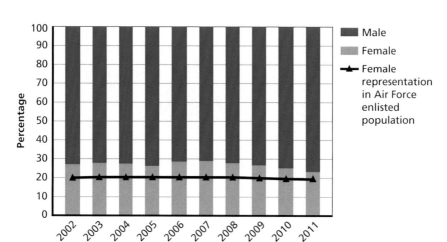

SOURCE: Analysis of DMDC data on active-duty Air Force enlisted personnel
(FY 2002–2011).
RAND RR1008-3.4

2005–2008) and afterward (FY 2009–2011). As Figure 3.3 showed, the gender retention gap existed during and after the drawdown period. The size of the gap did not change much, either. However, the gender gap is narrower for personnel with less than six years of service than with more than six years of service. The narrower gender gap for personnel with less than six years of experience maps onto gender trends in reenlistment in the 2000s. Specifically, the gender gap in Air Force reenlistment rates was narrower for first-term reenlistment (Zone A) than for later terms of reenlistment (Zones B and C) in the Air Force between 2000 and 2008. Moreover, Air Force women had higher first-term reenlistment rates than Air Force men from 2003 to 2006 (Military Leadership Diversity Commission, 2011a). Prior research on first-term attrition of enlisted service members suggests that women have higher first-term attrition than men, but studies show variations across services.

Figure 3.6 provides the adjusted female CCRs for the drawdown era, compared to the observed (unadjusted) female and male CCRs for

Figure 3.5
Air Force Enlisted CCRs, by Gender and Era, FY 2001–2011

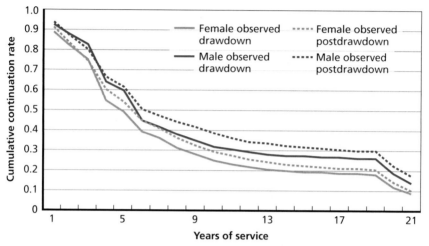

SOURCE: Analysis of DMDC data on active-duty Air Force enlisted personnel
(FY 2001–2011).

RAND RR1008-3.5

the drawdown era.[9] As with the Army officer adjustments in Chapter
Two, the adjustment to Air Force enlisted female CCRs in the 2000s
drawdown did not help explain all of the gender difference in retention
rates. This suggests that force structure changes during the drawdown
that may have affected the compositions of the male and female populations cannot fully explain the gender difference in retention rates.
Other factors we did not include in the model may help fill some of
the gap (e.g., parental status). For example, if enlisted women have
more family responsibilities (e.g., child care) than male peers, part of
the gender gap widening after five YOS could be partly a function of

---

[9] We also examined the postdrawdown gender gap in retention. The pattern was similar
to what we found for the 2000s drawdown era. The main difference between eras is that the
gender gap is slightly larger for the postdrawdown era, and the adjusted female CCR line
crosses the observed male CCR line at eight YOS instead of at seven YOS as in the drawdown
era. Because of the similarity in patterns across eras and because the gender gaps are not so
large that it would be difficult to distinguish all of the lines on a chart, we do not to show the
postdrawdown results in a figure.

Figure 3.6
Actual and Adjusted CCRs, by Gender, in 2000s Drawdown Era: Air Force Enlisted

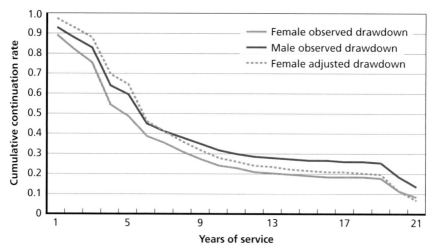

SOURCE: Analysis of DMDC data on active-duty Air Force enlisted personnel
(FY 2001–2011).
RAND RR1008-3.6

the gender difference in family responsibilities. The gender retention gap may also relate to gender differences in workplace experiences (e.g., gaining acceptance). For example, a 2001 RAND study on female and minority officer retention found that female officers in focus groups reported concerns with gaining recognition and acceptance from male peers (Hosek et al., 2001). The same study also found that female officers reported having more child care and household responsibility issues than their spouses. Unfortunately, most studies on women's retention focuses on officers, not enlisted personnel.

Summary

The 2000s did not bring the same general increase in demographic diversity as happened in the 1990s. The Air Force enlisted force, in particular, lost some demographic diversity between FY 2001 and 2011, in the middle of which occurred the Air Force drawdown. In this chapter, we unpacked gender diversity trends in the Air Force enlisted force, showing that a reduction in female shares of accessions starting in the mid-2000s contributed to female representation loss. Some of this loss may have been a function of Air Force accession cuts and tightening of entry standards.

Like the Army officers in the 1990s, female enlisted personnel in the Air Force in the 2000s had lower retention rates than males. We show that the gender retention gap existed throughout the decade (before, during, and after the drawdown) and could not be explained away by adjusting female retention rates to look like male retention rates in terms of other workforce characteristics, such as occupation, rank, and time in grade.

Force changes other than demographic diversity occurred during and after the Navy and Air Force reductions in the mid-2000s. Unlike its 1990s reductions, the Navy's mid-2000s reductions led to a more junior force and an officer corps leaning more toward tactical operations occupations. Unlike the Navy, the Air Force retained its enlisted-officer balance but became more heavily oriented to tactical operations than the Navy, especially in its enlisted force. The Air Force officer corps also became more senior, as it did after the 1990s drawdown period.

CHAPTER FOUR

Law, Policy, and Plans for Recent Active-Duty Drawdowns

As in the drawdown of the 1990s, the services have been using a variety of tools to reduce and shape their forces in recent drawdowns. Congress authorizes many tools through NDAA legislation and Title 10 authorities. These authorities are translated into DoD and service policy. The services also tailor tools taken from these authorities to achieve their own force reduction goals. In this chapter, we describe what we learned from service and OSD experts about recent plans and strategies for reducing the active-duty military forces in the Army, Marine Corps, and Air Force.

Discussions with Experts on Recent Drawdowns

We held discussions with service manpower and personnel policy experts to identify the services' general goals and strategies for upcoming active force reductions. The one exception is the Navy, which recently completed a long drawdown and is not planning to reduce its active force further. Since some of these experts were also able to provide information on the services' past reductions, we asked them the same types of questions about the current or upcoming reductions that we asked about past reductions. The interview questions for past and current reductions are in Appendix C.

To understand the drawdown tools currently available for use by the services, we also interviewed policy experts in the Office of the Secretary of Defense (Personnel and Readiness) [OSD(P&R)] office of Military Personnel Policy (MPP) in fall 2013. These experts provided an overview of available tools and information regarding service applicability, whether a tool focused on involuntary or voluntary separations, or if its primary use was for the officer or enlisted ranks. Additionally, we reviewed publicly available sources to augment the expert interviews and information provided by MPP. These sources included online service press releases and other relevant news articles (see Appendix C for methodology). We also reviewed relevant laws and policy documents to identify drawdown tools and associated authorities. We provide details on the specifications of sample drawdown tools available to all services (e.g., SERB) and sample tools developed by certain services. Our review of current drawdown tools is in Appendix D.

Size of Recent Reductions

Except for the Navy, the services are reducing their active-duty forces between FY 2012 and FY 2019. However, the magnitude of the reductions is a moving target. The Army is a good example of changes to drawdown targets. In January 2011, Secretary of Defense Robert Gates and Chairman of the Joint Chiefs of Staff Admiral Mike Mullen announced active Army end-strength would be reduced by 27,000 troops starting in 2015. In January 2012, DoD announced that the Army's end-strength would go from a high of 570,000 in 2010 to 490,000 in 2017 (Feickert, 2014a). However, the sequestration cuts that began in March 2013 affected the Army's plans, with the Army Chief of Staff, General Raymond T. Odierno, telling *Defense News* in the fall of 2013 that if sequestration remained in place, the Army would have no choice but to fall below the 490,000-soldier threshold it had set. In the meantime, the Army would accelerate its drawdown, coming down to 490,000 by 2015 instead of 2017. It would also potentially reduce 25 percent of personnel assigned to headquarters led by two-star or higher ranked generals (or civilian equivalents), including Army Forces Command, Training and Doctrine Command, and Army service component commands (Feickert, 2014a). By January 2014, the Army adjusted its plans, with General Odierno arguing that 450,000 was the minimum number of active-duty troops needed to implement the new defense strategic guidance. Army leadership also indicated some willingness to consider reducing active-duty personnel to 420,000 by FY 2019, but only if the Army National Guard downsized from 354,000 to 315,000 and the Army Reserve from 205,000 to 185,000 ("DoD Makes It Official: Budget Cuts Will Shrink Army to 420,000 Soldiers," 2014).

Estimates of the Marine Corps reduction have also changed over time. In January 2012, DoD announced the Marine Corps would reduce from a high of 202,000 to 182,000 by 2017. By April 2013, Secretary of Defense Chuck Hagel announced the Marine Corps could diminish from 182,000 marines to anywhere between 150,000 and 175,000 marines. In November 2013, then–Commandant of the Marine Corps General Amos testified to the Senate Armed Services Committee that 174,000 marines would be the lowest end-strength the Marine Corps would aim to achieve (Feickert, 2014b). However, according to the Marine Corps experts we interviewed in spring of 2014, the Marine Corps active end-strength was expected to reduce to 175,000 by the end of FY 2017, 1,000 more marines than estimated the previous year. In fall of 2014, a Marine Corps policy expert indicated the original 182,000 could hold.

Because the Air Force had already reduced its force in the past several years, the Air Force active-duty reductions will be a smaller percentage of the force than for Army and Marine Corps. Authorized active end-strength for the Air Force contracted from 332,800 in the FY 2012 NDAA to 327,600 in the FY 2014 NDAA, a loss of 5,200. As of FY 2014, the Air Force projected to reduce its active end-strength to about 308,037 by FY 2017 and 306,620 by FY 2019. This would result in a nearly 8 percent reduction of the Air Force active force between FY 2012 and FY 2019.

Service Goals and Strategies

Many of the goals from past reductions still hold for the recent reductions to the active military force. The primary motivation for a drawdown is to reduce military budgets, which the services achieve by reducing end-strength and force structure. Beyond this budgetary goal, the services try to retain their top performers and personnel in critical skill areas. The services do not have diversity goals associated with the drawdown; service policy experts argue that they base drawdown decisions on merit, not demographics. At most, some services consider how some of their drawdown decisions could affect different demographic groups but do not use the information to shape drawdown decisions. Based on our interviews, the legal fallout from the affirmative action lawsuits regarding the officer retention boards in the 1990s drawdown is a likely contributor to the reluctance of the services to consider anything related to diversity or equal opportunity when making drawdown decisions.

The general strategies of each service involve tools similar to those available in earlier drawdowns, although perhaps with added features. Combined, the tools allow the services to target personnel based on experience (e.g., YOS, rank, TIG), occupational category, and quality (e.g., performance, discipline). Although these three policy themes existed during the 1990s, the current reductions place less emphasis on accession cuts and more emphasis on separations of personnel considered lower in merit or quality than their peers and/or in overage skill areas (i.e., areas with more personnel than needed and considered less critical to the service mission than other areas). In the following sections, we describe the general service strategies for active military reductions for FY 2012 and beyond.

Army

The goals expressed for the current Army drawdown do not differ substantially from the objectives stated during the post–Cold War drawdown. In general, the Army intends to reduce the size of its force while maintaining readiness and caring for soldiers and families as they depart the service. According to experts we interviewed, those charged with implementing the Army's drawdown were directed to (1) ensure reductions are measured, not precipitous, and (2) maintain a balanced force by focusing on overage specialties. Another purported goal for Army reductions is to protect the institutional army from significant cuts so that it may be utilized as a base for rebuilding the force in the event of a future ground war (Feickert, 2014a).

According to experts we interviewed in fall 2013, the Army does not expect to reduce its force only through natural attrition. To achieve end-strength cuts, the Army will cut accessions as it did in the 1990s. Unlike the 1990s, the Army plans to rely on involuntary separations to achieve the remainder of its personnel reductions. According to G-1 (Army personnel policy) officials, the Army does not have the funds for voluntary measures. Moreover, at the time of our interviews, G-1 officials indicated that the Army expected that the poor U.S. economy would make it prohibitively expensive to entice Army personnel to leave voluntarily. Some of the involuntary measures put tighter controls on important military career stages such as reenlistment and promotion. For example, the experts we interviewed stated that the Army plans to reduce retention points for promotable specialists and corporals and promotable sergeants from 12 to eight years and 15 to 14 years, respectively. On the officer side, the Army has been making "up-or-out" policies more restrictive by targeting for separation majors passed over twice for promotion and lieutenant colonels with more than 20 years of service.

The Army is also using selection boards to involuntarily separate or retire personnel. According to the experts we interviewed, there is a perception among Army leadership that the voluntary separation measures of the 1990s induced too many high-quality personnel to leave the service. The Army therefore wants more control of the type of personnel who leave. For enlisted personnel, the Army is using different types of boards to target personnel for separation. One type of board-based program is the Qualitative Service Program, which separates personnel (mostly noncommissioned officers) with lower job performance, who are not promotable, and/or in overage specialties (Wiggins, 2012). For officers, most of those targeted for separation will be in year groups when the Army was growing the force in the mid-2000s. Many of these officers were captains as of FY 2013. However, the Army will also target some majors to avoid cutting too much from captains. The Army hopes to accomplish these types of reductions by removing personnel from the bottom of each branch in terms of quality.

The Army also plans to use some traditional involuntary measures, such as RIFs, SERBs, and early retirement authorities. Specifics on some of these measures can be found in Appendix D.

Marine Corps

According to the Marine Corps experts we interviewed, the Marine Corps Commandant gave the following drawdown guidance to the Manpower Plans Division in 2011: (1) "keep the faith" with Marines, but if Marines are not "giving their best," the Corps is not "obligated to keep faith with them"; and (2) use traditional attrition measures. Because the Marine Corps is a youthful service—almost half of the service consists of corporals and below—it experiences a high rate of turnover even in normal times. Specifically, about 30,000 to 35,000 Marines leave the active force each year. The reduction of approximately 5,000 Marines a year in the current drawdown is not a particularly large addition to the service's normal attrition. Consequently, Marine Corps manpower officials do not think they need to take extraordinary measures to induce people to leave.

Nonetheless, the Marine Corps manpower experts we interviewed recognize that the whole "manpower pyramid" must shift to the left to correctly execute a drawdown. The plan is to rely less on accession cuts than the Marine Corps did in the 1990s, because the lesser opportunity at higher grades resulted in promotion stagnation, among other adverse outcomes. The lesson learned is to decrease accessions while also making use of traditional tools to prevent an imbalance in the higher reaches of the "pyramid." The Marine Corps also plans to balance the force by skill level, cutting more from infantry and artillery battalions and from aviation squadrons than from key skill areas such as special operations and cyber (Feickert, 2014b).

A centerpiece of the Marine Corps separation strategy is controls on reenlistment. In the past, Marines would reenlist on a first-come, first-served basis. As long as the Marine was upwardly mobile (i.e., able to be promoted), he/she could have a career in the Marine Corps. However, in May 2011, the Marine Corps issued Marine Administrative Message 273/11 to announce the move to a merit-based tiered system for determining eligibility for reenlistment. Marines applying to reenlist would be placed into one of four tiers based on a variety of quality indicators such as their physical fitness test scores, performance and conduct marks, rifle range scores, and meritorious promotions.[1] Marines in the first tier would have their reenlistment packages processed more quickly than Marines in the lower tiers.

In addition to the tiered evaluation system to control reenlistments, the Marine Corps introduced "Quality Marine" waivers to allow commanders to reenlist "hard-charging" Marines in their specialty even if there are no "boat spaces left" (Lamothe, 2011). In addition, the Marine Corps advised career planners and commanders to recruit high-performing Marines from nontechnical, overage specialties into critical technical specialties such as cyber and intelligence (Lamothe, 2011).

Beyond reenlistment controls, the Marine Corps planned to use TERA for Marines with 15 or more YOS and VSP for midgrade officer and enlisted personnel with six to 15 YOS. The Marine Corps also planned to use SERBs for lieutenant colonels and colonels. After SERBs began, the Marine Corps saw an uptick in officers retiring voluntarily. The voluntary retirements helped the Marine Corps avoid heavier reliance on SERBs.

Air Force

Although the Air Force reduced its active force between FY 2005 and 2008, the Air Force had continued its force-shaping efforts after FY 2008. According to one Air Force expert in FY 2014, less than a quarter of the cuts to Air Force end-strength in FY 2015 would be expected to come from cutting accessions. The rest of the reduction in FY 2015 would have come from separations, natural or induced via drawdown measures. The Air Force's general strategy to induce separations is to maximize voluntary separations by offering incentives. However, the Air Force targets those incentives at personnel it would otherwise

---

[1]   Physical fitness tests and rifle range tests are likely more valid measures of abilities and skills for infantry and other physically demanding combat fields than for technical fields, such as cyber. The description of the tiered evaluation system does not specify how the Marine Corps weighs physical fitness test scores and rifle range test scores when evaluating Marines from technical fields versus Marines from combat fields.

involuntarily separate: personnel with lower levels of performance, who are not promotable, and/or in overage skill areas. For example, the Air Force offered a voluntary incentive, VSP, in 2010 prior to conducting RIF boards for eligible officers with six to 12 YOS (Parcell, 2011).

Like Army experts we interviewed, Air Force experts cite concerns that too many high-quality personnel will leave if separation programs are not targeted toward lower-quality personnel. To address this concern, the Air Force is using different approaches to separate personnel. One approach used on enlisted personnel is enlisted discharge authorities. The Air Force created the DOS Rollback program, which accelerates the DOS of enlisted personnel up through E-8 grades with specific reenlistment eligibility codes. These codes include declining a permanent change of station, temporary duty, training, retraining, or professional military education (Losey, 2013a). Other codes focus on negative quality indicators like serving punishment for an Article 15 violation.

In addition to reenlistment controls, the Air Force uses board processes to involuntarily separate enlisted personnel. Boards vary in whom they target; personnel are targeted based on a combination of grade, years of service, and specialty. For example, Enlisted Retention Boards target enlisted personnel from senior Airmen (E-4) through senior master sergeants (E-8) in overage specialties (Losey, 2013c), whereas Chief Master Sergeant Retention Boards target Chief Master Sergeants with 20 years of total active federal military service and in overage specialties (Gildea, 2013).

Another type of enlisted retention board that the Air Force is using is the Quality Force Review Board. According to an expert we interviewed, the board reviews individual records to determine whom to retain. The records will include "retention recommendation forms" (RRFs), which have recommendation categories akin to those used in promotions (i.e., a commander recommends "definitely retain," "retain," or "do not retain").[2] The Quality Force Review Boards apply to personnel with less than 18 years or more than 20 years of ser-

---

[2]   RRFs are also used in officer separation boards such as RIFs, SERBs, and Force Shaping Boards (FSBs).

vice who have "specific negative reporting identifiers, reenlistment eligibility codes, assignment availability codes or grade status reasons" (Gildea, 2014). These negative identifiers reflect performance and conduct issues, including Article 15s, absenteeism, and limited/lack of promotability (i.e., at a rank or skill level not commensurate with the grade) (Losey, 2013b). Involuntary separation measures like these are generally accompanied by some form of separation or retirement compensation for eligible personnel. For example, enlisted personnel with six to 15 years of active service will receive full separation pay (Gildea, 2014). According to Losey (2014a), the Air Force requested TERA for up to 4,200 enlisted personnel for FY 2015.

The Air Force also uses board processes to involuntarily separate officers. Programs include Force Separation Boards (FSBs) for officers with three to six YOS, RIFs, and SERBs, and Enhanced SERBs (E-SERBs). E-SERBs provide more flexibility than traditional SERBs by allowing the service to target officers by occupational category and in lower grades than O-5. E-SERBs also allow the same officers to be reviewed annually, as opposed to every five years per a traditional SERB. (See Appendix D for more details on SERBs and E-SERBs.) Officers subject to SERBs, E-SERBs, or RIFs are eligible to apply for VSP or TERA, depending on their YOS.

Summary

As in the 1990s, the Army, Marine Corps, and Air Force have been experiencing another major reduction in the sizes of their active-duty forces. These reductions are planned through FY 2019, with the Army taking the largest hit. To achieve their reductions, the Army, Marine Corps, and Air Force have been using a variety of tools and programs with an overarching goal of balancing their forces by focusing cuts on overage occupational categories and experience categories (e.g., officer year groups), and on lower-quality personnel. Unlike the 1990s drawdown, the recent drawdowns are expected to rely less on accession cuts and more on separation measures, including involuntary measures.

In the next chapter, we describe how we use some of the drawdown programs described in this chapter to develop and analyze notional drawdown scenarios. We use these scenarios to examine how cuts based on occupational category, experience level, and, in one case, quality could affect women and minority groups.

CHAPTER FIVE

Potential Impact of Recent Drawdowns on Demographic Diversity in Active-Duty Force

This chapter offers an overview of the methodology, findings, and policy implications of potential drawdown impacts on women and racial/ethnic minorities. We developed and analyzed scenarios representing different drawdown strategies that vary along one or more drawdown-strategy themes identified in Chapter Four—occupation, experience, and quality. To develop these scenarios, we culled information from our interviews and media sources to identify different types of drawdown programs that the services used in the recent past or reportedly may use in upcoming reductions. Using FY 2012 personnel data, we analyzed demographic group differences in sizes of force cuts based on each scenario. These analyses point to the potential adverse impact of certain drawdown strategies. After our analysis, we met with force management experts and diversity policy experts to discuss potential policy options for addressing demographic diversity in a drawdown context.

Scenario Development and Analytic Strategy

Scenario Development

As with our historical drawdown analyses, our scenario analyses are based on data from the DMDC Active Duty Master File. We examined how reductions to each service's FY 2012 active force would affect force reductions (or "cuts") to female personnel, black personnel, and Hispanic personnel. We used FY 2012 as our benchmark because it is the last year for which we have complete data.

We developed notional scenarios to reflect different types of drawdown programs that the services have used within the past few years or may use within the next year or two (through FY 2015). To develop scenarios, we pulled details from interviews with service drawdown experts and from publicly available information, namely news articles or the services' public websites. Because the Navy did not plan an active force drawdown for FY 2012, we built Navy scenarios based on drawdown programs used in the Navy's mid-2000s drawdown. For each service, we created scenarios that target different subpopulations based on corps (enlisted and commissioned officers) and experience levels (namely, grade/rank).

Our main challenge in building scenarios was finding enough details about drawdown programs. For example, a service website might note that specific occupational groups will be targeted without listing the occupational groups. In general, we did not have enough details on which occupations or experience levels to target. To address this issue, we took liberties in varying features of the scenarios, particularly with occupational cuts. Most occupational comparisons involved tactical operations (ops) occupations versus nontactical ops occupations.[1] We chose this occupational comparison for two reasons: (1) prior research shows that women and racial/ethnic minorities are underrepresented in tactical operations occupations, and (2) senior leaders tend to come from tactical operations occupations.[2] Tables 5.1 and 5.2 show the percentage of white men in enlisted and officer occupational categories by service, corps, and rank category. We shaded cells gray if the percentage of men in a given occupational category and rank group was higher than the percentage of men in the rank group overall (total). Across services and corps, men are overrepresented in tactical operations occupations and underrepresented in health care occupations.

To examine the sensitivity of the scenarios to the size of cuts, we varied cut sizes. We did our best to use reported cut sizes as anchors

---

[1]   To save space, we refer to tactical ops as "tactical" and nontactical ops as "nontactical" in tables in the following sections.

[2]   For a detailed discussion about the relationship between tactical operations occupations and senior military leadership, see MLDC (2011b).

Table 5.1
Percentage Representation of White Men in Active-Duty Enlisted Occupational Categories, by Service (FY 2012)

Occupational Category	Army E1–E4	Army E5–E6	Army E7–E9	Navy E1–E4	Navy E5–E6	Navy E7–E9	Marine Corps E1–E4	Marine Corps E5–E6	Marine Corps E7–E9	Air Force E1–E4	Air Force E5–E6	Air Force E7–E9
Communications and intelligence	64.4	66.6	55.7	35.3	40.6	54.3	66.4	60.2	61.0	61.6	59.3	65.1
Craftsworkers	51.6	51.3	43.0	37.9	56.1	61.6	68.6	54.9	51.8	68.0	60.6	72.9
Electric/mechanical equipment repair	58.6	59.1	48.9	35.4	47.8	56.2	68.6	64.7	63.1	72.1	71.1	78.8
Electronic equipment repair	53.0	52.8	45.7	42.4	52.3	67.5	69.4	65.4	61.2	69.6	65.5	71.1
Functional support and administration	24.5	27.6	30.8	22.8	24.4	35.2	37.7	35.4	41.6	43.9	34.4	42.6
Health care	45.7	42.4	35.5	31.4	29.1	46.9	--	--	--	32.2	27.2	37.0
Infantry, gun crews, and seamanship (tactical ops)	72.7	71.4	64.7	38.0	53.9	68.7	78.4	72.0	69.0	73.5	76.7	81.5
Other technical and allied specialists	51.5	53.3	50.9	52.2	65.9	78.8	71.2	68.9	68.5	67.8	67.3	69.8
Service and supply handlers	44.2	45.6	40.5	27.4	35.3	40.0	57.7	46.7	43.9	58.0	57.2	64.3
Total (baseline)	55.6	54.5	48.7	34.9	44.3	55.9	66.1	59.0	55.1	60.4	56.3	62.1

SOURCE: Analysis of DMDC data on active-duty enlisted personnel (FY 2012).
NOTES: Occupational categories are based on those used by DMDC to compare occupational groups across services. The Marine Corps does not have health care personnel, hence the dashes for those cells.

Table 5.2
Percentage Representation of White Men in Active-Duty Officer
Occupational Categories, by Service (FY 2012)

	Army		Navy		Marine Corps		Air Force	
Occupational Category	O1–O3	O4–O6	O1–O3	O4–O6	O1–O3	O4–O6	O1–O3	O4–O6
Administrators	41.6	52.9	45.5	50.6	54.2	57.6	45.3	62.0
Engineering and maintenance officers	59.4	64.8	65.6	73.3	67.0	69.3	66.0	72.2
Health care officers	39.6	54.3	37.0	53.4	–	–	34.8	52.5
Intelligence officers	57.7	68.0	60.1	67.9	70.7	81.5	53.2	68.4
Scientists and professionals	64.6	72.8	62.3	69.7	66.8	74.6	61.9	74.2
Supply, procurement and allied officers	48.5	57.3	52.5	61.5	60.6	67.0	53.3	68.5
Tactical operations officers	78.2	80.8	70.1	79.8	78.6	85.2	77.3	84.5
Total (baseline)	59.8	66.4	60.7	68.6	71.0	79.4	62.8	72.3

SOURCE: Analysis of DMDC data on active-duty officers (FY 2012).

NOTES: We did not list the "general officers and executives" category because it is made up largely of the majority of the generals and admirals (i.e., those in O7–O10 ranks). Likewise, we did not report the O7–O10 ranks, since all officers in those ranks fall into the "general officers and executives" occupational category. The Marine Corps does not have health care personnel, hence the dashes for those cells.

for the scenarios,[3] but cut sizes were not always available, or they were subject to change.[4]

---

[3] Although we did not explicitly examine the impact of current economic conditions (e.g., U.S. unemployment rate) on the scenario outcomes, our selection of cuts that roughly represent the size of cuts proposed or used by the services reflects, to some degree, the services' estimate as to the size of cuts needed given external (economic, social) and internal service conditions.

[4] We find that varying cut sizes did not affect the results much, if at all. In tables, we display results for the midsize cuts. The midsize cuts generally reflect the estimated cuts reported in our sources.

Another challenge lies with our personnel data. Because we have occupational category and experience measures (e.g., YOS) in the data, we were able to vary scenarios based on occupation and experience. However, the DMDC data do not include much in the way of personnel "quality" information. The only quality measure we had and that can be used in a drawdown decision is AFQT score. We used AFQT to create a scenario whereby accession cuts are made while AFQT standards are tightened (i.e., recruits need higher AFQT scores to enter). We applied this scenario to each service. We varied the size of accession cuts (10-percent, 20-percent, or 30-percent cut to a service's FY 2012 accessions) and applied all cuts to the lowest AFQT categories, Category IIIA and Categories IIIB–IV, as we assume that the services would wish to maximize accessions among higher AFQT categories (I and II).

Analytic Strategy

Since many drawdown decisions ultimately rely on a person's performance records, we could not estimate how the services actually select personnel for separations. Without this information, we instead assigned cuts randomly across the target population(s) and baseline population for each scenario. The cuts are therefore proportional to the size of the demographic groups in each population. In reality, selection processes are not random; systematic differences between demographic groups in terms of career fields, experience levels, and quality factors likely result in nonrandom selection effects. Therefore, our results could change were nonrandom selection factors included in the analysis.

To describe the results, we compare cuts from a targeted population to cuts from a baseline population, by demographic group: women, non-Hispanic blacks, and Hispanics.[5] If the proportion of cuts to a demographic group is higher in the target population than in the baseline population, the drawdown program has potential adverse

---

[5]   In a couple of scenarios, we did not have a reasonable cut size to target. In such cases, we just compared the demographic group distribution in the target population(s) to the demographic group distribution in the baseline population. In our scenario description tables, we include a column on cut sizes. Scenarios not based on specific cut sizes have an "N/A" or "not applicable" comment.

results for that demographic group. For example, assume a separation program cuts 100 people from a target population, and 20 of those 100 are women (20-percent female). However, a cut of 100 people from the baseline population results in 10 women being cut (10-percent female). The 20-percent cut from the target population is larger than the 10-percent cut from the baseline population. Thus, the drawdown program could have an adverse impact on women in that baseline population. In the opposite case (baseline cuts higher than target cuts), we do not have evidence of potential adverse impact.

In our results tables, we provide the percentages for each demographic group in the targeted (scenario) population and baseline population (e.g., for women, 20 percent for the scenario and 10 percent for the baseline in the previous example); we do not report differences between scenario and baseline percentages. For a few scenarios, we did not have specific cut sizes to use, so we instead compared the demographic group representation (percentages) in the targeted population to the baseline population. To aid readers in identifying differences between baseline and scenario (targeted group) percentages, we use different shades of gray to represent cases of potential adverse impact[6] (dark gray) and cases of no adverse impact (light gray). Please note that the gray shading reflects demographic group comparisons, not comparisons in force capability or personnel quality. Therefore, findings of adverse impact (or even nonadverse impact) would need to be considered in the context of the services' other drawdown policy goals (e.g., maintaining certain force capabilities).

Race/Ethnicity Breakouts by Gender

Because of the sheer number of demographic groups, services, and corps, we did not explore the interactions between gender and race/ethnicity for our main scenario analyses. However, drawdown decisions could conceivably affect minority men and women differ-

---

[6] Because we do not have all of the measures used by the services to select personnel for drawdown programs, we do not conduct typical adverse impact analyses. Adding other relevant measures could also change the results, so we use terms like "potential" and "possible" to describe our findings. For more details on how adverse impact is typically modeled and concerns with that modeling, see Roth, Bobko, and Switzer (2006).

ently than white men and women. To explore whether race/ethnicity interactions with gender would change our results, we analyzed a select number of scenarios for the following demographic groups: non-Hispanic white women, non-Hispanic black women, Hispanic women, non-Hispanic black men, and Hispanic men. To select scenarios, we started by picking one officer scenario and one enlisted scenario from each service. Within each service-corps scenario set, we chose a scenario with the largest or one of the largest targeted population sizes and then examined the demographic group sizes within those targeted populations to determine whether there were enough individuals in each group to conduct the analysis. Although there is no simple rule to determine sufficient group size, we chose to do a gender-by-race/ethnic analysis if most of the demographic groups for that scenario had at least 50 individuals.[7] Based on this rule, we determined that we could not analyze any Marine Corps officer scenarios because there were too few female officers to break out into race/ethnic groups. For scenarios where we perform these analyses, we describe them later in the chapter among our main scenario analyses.

Army

We analyzed five Army drawdown scenarios. Two scenarios cover officers: (1) RIFs of captains and majors with YOS and occupational

---

[7] We used a Poisson approximation to the hypergeometric distribution function to estimate the probability that at least five individuals from a demographic subgroup would be selected at random for cuts, given a certain cut size, targeted population size, and demographic group size. For example, assume a targeted population of 20,000 people, 150 of whom are in a particular demographic group. Assume there will be a cut of 500 people from the targeted population of 20,000 people. There is approximately an 82 percent chance that five or fewer people from the demographic group of 150 people would be selected at random for cuts. Our calculations show that a demographic group with fewer than 50 people across a range of targeted population and cut sizes was associated with at least a 60-percent chance of five or fewer people being selected for cuts. Population sizes were too small to break out analyses by gender, race, and ethnicity (black Hispanic women, white Hispanic women, etc.).

variations,[8] and (2) SERBs of lieutenant colonels and colonels with TIG variations. The next three scenarios cover enlisted: (3) accessions cut with tightening of AFQT requirements; (4) reduced retention control points for E-4s and E-5s with occupational variations; and (5) involuntary separations (as in an enlisted board process) of E-7s through E-9s with occupational variations. Table 5.3 gives details on scenario features.

Officers

Table 5.4 provides results for the two officer scenarios. Scenario 1a shows that a RIF of captains with four to six YOS might not adversely affect women, blacks, and Hispanics. Instead, a RIF that focuses on all YOS could have more of an adverse result on these demographic groups than a RIF focused only on captains in the four-to-six-YOS band. A RIF of majors, however, has the potential for adverse impact against women and Hispanics. If the Army were to target all occupational categories for majors with nine to 13 YOS, 16.9 percent of cuts would come from women compared to 15.9 percent if cuts were not restricted to the nine-to-13-YOS group (baseline population). More explicitly, about 93 of the 550 majors with nine to 13 YOS across all Army officer occupational groups targeted for cuts could be expected to be women. If instead the 550 cuts came from the entire population of majors (baseline population), 87 women could be cut. Therefore, a cut of 550 majors from the scenario population could result in six more female majors (93 − 87 = 6) being cut than if the cuts instead came from the whole population of majors.

---

[8] The occupational variations in this case refer to the Army Competitive Category (ACC). According to Gibson (2007, p. 2), the ACC "consists of all Army branches and career fields except specialty branches (such as the Medical Service Corps, the Veterinary Corps, Chaplains Corps, etc.)." We approximate the ACC by removing the DoD occupational categories of "Health Care Officer" and "Scientists and Professionals." Although our two Army officer scenarios focus on the ACC, we focus on occupational variations (i.e., ACC versus all occupations) for the RIF scenario but do not vary occupation for the SERB scenario (i.e., all ACC). Because a defining feature of a SERB is its focus on officers with longer TIG, we wanted to examine whether TIG variations (controlling for occupational category) would affect the results.

A reason for the potential adverse impact of the RIF of majors is that female majors are disproportionately represented among the health care occupational group. Opening the RIF up to health care occupations could therefore put more female majors at risk of separation. Hispanic majors could be adversely affected in either scenario variation, suggesting that Hispanic representation among the nine-to-13-YOS group overrides the occupational factors we examine in the scenario.

Although the two RIFs do not show potential adverse impact against black officers, the same does not apply to the SERB scenario. A SERB of lieutenant colonels or colonels may not bode well for black officers because SERBs focus on officers with long TIG or who have been passed over for promotion. Black officers are overrepresented among lieutenant colonels and colonels with long TIG, making a SERB more negative for black officers than officers in other racial/ethnic groups. Conversely, these types of SERBs may not adversely affect the Army's female officers and Hispanic officers.

Table 5.5 provides the gender-by-race/ethnicity breakouts for scenario 1. When women are broken out into racial/ethnic groups, we find that non-Hispanic white women who are captains with four to six YOS (scenario 1a) could be adversely affected by RIFs if those RIFs target all occupational groups. As we found for the main analysis of scenario 1a, RIFs of captains with four to six YOS do not adversely affect any of the other minority demographic groups. Scenario 1b, RIF of Army majors with nine to 13 YOS, also shows different patterns by demographic group when women and men are broken out into different racial/ethnic categories. Like the RIF of captains, the RIF of majors could adversely affect non-Hispanic white women if the RIF targets all occupational groups, not just ACC occupations. Unlike the main analysis of scenario 1b, we find evidence of potential adverse impact for non-Hispanic black personnel, but focused on men in ACC occupations. Taken together, the findings demonstrate how Army RIFs could have different patterns of adverse effects depending on the gender and racial/ethnic identities of those being targeted for reductions.

Table 5.3
Details for Army Drawdown Scenarios

Scenario Number	Program	Targeted Population	Scenario Variation(s)	Baseline Population	Cut Sizes
1a*	RIF[a]	Captains with 4–6 YOS	• Proxy ACC occupations • All occupations	All captains	• 600 • 1,200 • 1,800
1b*	RIF[a]	Majors with 9–13 YOS	See cell above	All majors	• 200 • 550 • 800
2a	SERB[b]	Lieutenant colonels with ≥ 4 years TIG	• ≥ 4 years TIG in proxy ACC occupations • Any TIG in proxy ACC occupations	All lieutenant colonels	• 30% of targeted population (1,953)
2b	SERB[b]	Colonels with ≥ 4 years TIG	See cell above	All colonels	• 30% of targeted population (783)
3	Accessions cut and AFQT requirements tightened	Accessions (0–1 YOS)	• All cuts from Categories IIIB–IV (100%) • 75% Categories IIIB–IV, 25% Category IIIA • 50% Categories IIIB–IV, 50% Category IIIA	All accessions	• 10% accessions (5,185) • 20% accessions (10,371) • 30% accessions (15,556)

Table 5.3—Continued

Scenario Number	Program	Targeted Population	Scenario Variation(s)	Baseline Population	Cut Sizes
4a	Reduced retention control points[b]	E-4s with ≥ 8 YOS	• ≥ 8 YOS in tactical • ≥ 8 YOS in nontactical • Any YOS in tactical • Any YOS in nontactical	All E-4s	N/A (results based on overall population proportions)
4b	Reduced retention control points[b]	E-5s with ≥ 14 YOS	• ≥ 14 YOS in tactical • ≥ 14 YOS in nontactical • Any YOS in tactical • Any YOS in nontactical	All E-5s	N/A (results based on overall population proportions)
5*	QSP or similar involuntary separation program[c]	E-7s, E-8s, and E-9s	• 75% of cuts from tactical, 25% from nontactical • 50% tactical, 50% nontactical • 25% tactical, 75% nontactical	All E-7s, E-8s, and E-9s	• 600 • 1,000

SOURCES: [a] Army Times (2014). [b] Interview notes. [c] Brown and Millham (2014).

NOTE: *Scenarios selected for race/ethnicity breakouts by gender.

Table 5.4
Main Results for Army Officer Scenarios

Scenario Number	Scenario Variations	Women	Non-Hispanic Black	Hispanic
1a	4–6 YOS and in proxy ACC only	12.6	7.5	6.7
	4–6 YOS and in any occupational group	18.8	7.6	6.1
	All captains (baseline)	19.8	13.7	7.7
1b	9–13 YOS and in proxy ACC only	12.8	13.4	6.6
	9–13 YOS and in any occupational group	16.9	12.7	6.3
	All majors (baseline)	15.9	14.1	4.3
2a	Long TIG (≥ 4 years) and in proxy ACC	8.0	13.8	5.2
	Any TIG and in proxy ACC	8.0	13.7	5.2
	All lieutenant colonels (baseline)	12.8	12.3	5.4
2b	Long TIG (≥ 4 years) and in proxy ACC	6.9	12.2	3.4
	Any TIG and in proxy ACC	6.9	12.0	3.5
	All colonels (baseline)	11.1	10.4	3.7

NOTES: Table values provide the percentage of cuts from each demographic group for the given scenario. Scenario 1 uses middle cut sizes in Table 5.3 (e.g., 550 for scenario 1b). Scenario 2 uses cuts in Table 5.3.

Enlisted

Table 5.6 shows results for Army enlisted scenarios. The AFQT scenario (No. 3) results show adverse impact potential for all three demographic groups. As cuts lean toward the lower AFQT categories, the percentages increase so that cuts heavily concentrated on Categories IIIB–IV recruits would have more adverse impact than cuts focused more on Category IIIA recruits.

Scenario 4—reduced retention control points—reveals an interaction of occupation and YOS with demographic group differences. If cuts focus on longer YOS (as per a retention control point reduc-

Table 5.5
Results for Army Officer Scenario with Gender-by-Race/Ethnicity Breakouts

Scenario Number	Scenario Variations	NH White Women	NH Black Men	NH Black Women	Hispanic Men	Hispanic Women
1a	4–6 YOS and in proxy ACC only	8.0	5.3	2.2	5.5	1.2
	4–6 YOS and in any occupational group	12.4	4.7	2.8	4.8	1.3
	All captains (baseline)	11.3	8.5	5.1	6.0	1.5
1b	9–13 YOS and in proxy ACC only	6.5	9.7	3.7	5.5	1.0
	9–13 YOS and in any occupational group	9.2	8.2	4.5	5.1	1.2
	All majors (baseline)	8.2	9.6	4.5	5.8	1.2

NOTES: Table values provide the percentage of cuts from each demographic group for the given scenario. Scenario 1 uses largest cut sizes in Table 5.3 (e.g., 800 for Scenario 1b). NH stands for "Non-Hispanic."

tion) and are heavier on nontactical operations occupations or split evenly tactical and nontactical, black E-4s and E-5s could be adversely affected. For example, 26.8 percent of E-4s with eight or more YOS and in nontactical operations are black. By comparison, only 19.3 percent of all E-4s are black. This suggests that black E-4s are more heavily concentrated in longer YOS (eight or more) and in nontactical operations occupations. Because 26.8 percent (scenario) is higher than 19.3 percent (baseline), we identify this variation of scenario 4a as one with potentially negative results for black E-4s.

However, if cuts are made irrespective of YOS, adverse impact against black personnel goes away for the even-split occupation variation. Moreover, the cuts are more severe for black personnel when YOS is restricted, suggesting that black E-4s and E-5s are more heavily represented among longer YOS groups.

In comparison, female E-4s and E-5s may not be adversely affected by reduced retention control points. If cuts instead focused on nontactical ops occupation but not reduced retention control points, women could face adverse impact. For Hispanics, the situation varies by grade.

Table 5.6
Main Results for Army Enlisted Scenarios

Scenario Number	Scenario Variations	Women	Non-Hispanic Black	Hispanic
3	100% cuts from Categories IIIB–IV	18.8	31.7	18.0
	75% Categories IIIB–IV, 25% Category IIIA	17.8	28.8	16.9
	50% Categories IIIB–IV, 50% Category IIIA	16.7	25.8	15.7
	All accessions (baseline)	14.7	20.2	13.2
4a	≥ 8 YOS and in any occupational group	11.0	25.0	12.9
	≥ 8 YOS and in nontactical group	12.4	26.8	12.7
	≥ 8 YOS and in tactical group	0.5	12.1	13.8
	Any YOS and in nontactical group	18.5	23.0	12.1
	Any YOS and in tactical group	1.1	8.1	11.8
	All E-4s (baseline)	14.2	19.3	12.8
4b	≥ 14 YOS and in any occupational group	9.4	34.5	11.0
	≥ 14 YOS and in nontactical group	10.0	35.7	10.8
	≥ 14 YOS and in tactical group	1.1	17.5	13.9
	Any YOS and in nontactical group	15.5	23.9	14.1
	Any YOS and in tactical group	0.8	8.3	12.4
	All E-5s (baseline)	12.2	20.3	13.7
5	75% tactical, 25% nontactical	4.0	21.6	12.5
	50% tactical, 50% nontactical	7.8	26.3	12.6
	25% tactical, 75% nontactical	11.6	31.0	12.7
	All E-7s, E-8s, and E-9s (baseline)	10.9	30.1	12.7

NOTES: Except for scenario 4, the table values provide the percentage of cuts from each demographic group for the given scenario. Scenario 3 assumes 20 percent of accessions cut. Scenario 4 is not based on specific cut sizes but reflects the demographic group proportions in the targeted and baseline populations. Scenario 5 uses a cut of 600 personnel.

For Hispanic E-4s, cuts based on longer YOS in tactical operations or even-split across the two occupational categories could have adverse impact, although the size of impact could be very small (1 percentage point or less). For Hispanic E-5s, cuts to tactical operations occupations among those with long YOS and cuts that ignore YOS but focus on nontactical ops occupations could have adverse impact against Hispanic E-5s. The results suggest that Hispanic E-5s with longer service are less concentrated in tactical operations than Hispanic E-5s with less time in service.

The last scenario (5) attempts to cut senior enlisted (E-7 through E-9) via a QSP board. Because we did not have performance data, we could not incorporate the quality element into the scenario. Instead, we use occupational variations to see how cuts to senior enlisted personnel would be affected by tactical versus nontactical cuts. If cuts focus heavily on nontactical operations (75 percent), adverse impact against all three demographic groups is possible.

The gender-by-race/ethnic group breakout results for the Army enlisted scenario 5 are presented in Table 5.7. As with the breakout analysis for Army officers, the breakout analysis for Army enlisted reveals differences from the main analysis of the scenario. In particu-

Table 5.7
Results for Army Enlisted Scenario with Gender-by-Race/Ethnicity Breakouts

Scenario Number	Scenario Variations	NH White Women	NH Black Men	NH Black Women	Hispanic Men	Hispanic Women
5	75% tactical, 25% nontactical	1.0	19.3	2.3	6.8	0.3
	50% tactical, 50% nontactical	1.9	21.8	4.5	6.7	0.6
	25% tactical, 75% nontactical	2.8	24.3	6.7	6.5	0.8
	All E-7s, E-8s, and E-9s (baseline)	2.7	23.8	6.3	6.6	0.8

NOTES: Table values provide the percentage of cuts from each demographic group for the given scenario. Scenario 5 uses largest cut sizes in Table 5.3 (i.e., 1,000). NH stands for "Non-Hispanic."

lar, breaking out Hispanics into male and female groups changes the findings. Although an involuntary separation of senior enlisted personnel does not show evidence of potential adverse impact against Hispanic enlisted women, there is evidence of potential adverse impact against Hispanic men if the cuts focus heavily on tactical operations occupations or split evenly between tactical and nontactical operations occupations. The findings suggest that Hispanic males in senior Army enlisted grades have relatively lower representation in nontactical operations occupations than in tactical operations occupations, so heavy cuts to tactical operations occupations can present an issue for them. This creates a dilemma for force programmers if some minority groups could be adversely affected by nontactical cuts but other minority groups could be adversely affected by tactical ops cuts.

Summary

Of the three main demographic groups examined, black personnel have the most cases of potential adverse impact. Scenarios with variations in cuts based on time in service or grade provide evidence that reductions based on longer service could disproportionately affect black personnel. Enlisted scenarios with occupation-based cuts (e.g., QSP [5]) suggest that heavy cuts to personnel in nontactical operations occupations could adversely affect black personnel. Our gender-by-race/ethnicity breakout analyses suggest that black male majors, but not black female majors, could also be adversely affected by cuts focused on ACC occupations. Thus, RIFs of majors in ACC occupations could be problematic for black men but not for other minority groups.

Like enlisted black personnel, enlisted female personnel could face adverse impact from heavy cuts to nontactical operations occupations. Moreover, white female captains and majors could be adversely affected by RIFs that include health care occupations and scientist and professional occupations (i.e., non-ACC occupations) mainly because of white female officers' overrepresentation in health care occupations. Such RIFs may not adversely affect female officers from minority race/ethnicity groups. In terms of cuts based on experience, cuts to enlisted personnel with longer YOS (scenario 4) could work to women's advantage in the enlisted force and officer corps.

No single theme dominates the findings for Hispanic personnel. For example, scenario 4a (reduced retention control points for E-4s) shows the potential of adverse impact for cuts based on long YOS and both occupational categories or long YOS and tactical operations occupations. In contrast, scenario 4b (reduced retention control points for E-5s) suggests adverse impact is possible for Hispanic E-5s with long YOS in tactical operations occupations or Hispanic E-5s with any YOS and in nontactical operations occupations. Our gender-by-race/ethnicity breakout analyses add to the complexity of the Hispanic findings. For example, the main analysis of scenario 5 (senior enlisted cuts) shows potential adverse impact for Hispanics if there are heavy nontactical occupation cuts. However, when Hispanic men are analyzed separately from Hispanic women, the trend for Hispanic men shifts toward heavier tactical cuts being adverse. Hispanic women show no adverse impact in this scenario. Overall, the findings for Hispanics in the Army suggest that occupational distributions of Hispanic personnel vary by year groups and grades, resulting in interactive effects.

The one scenario that has potential adverse impact for all three demographic groups is the AFQT scenario (3). Because women and minorities are overrepresented among lower AFQT categories, accession cuts focused on those categories would have a disproportionate impact on women and minorities. To the extent that the Army increased AFQT standards during an accession cut, demographic diversity of the junior Army enlisted force could be affected.

Marine Corps

We analyzed six Marine Corps scenarios. Three scenarios are for officers: (1) reduction to future career force (i.e., cuts to officers at four YOS), (2) Selective Continuation Board for majors with occupational variations, and (3) SERBs of lieutenant colonels and colonels with occupational variations (for lieutenant colonels only).[9] Another three

---

[9] We planned to compare tactical and nontactical operations occupations in the SERB scenario for colonels, but the Marine Corps classifies colonels in the occupational category

scenarios focus on enlisted: (4) accessions cut with tightening of AFQT requirements; (5) first-term reenlistment controls (i.e., reductions to personnel between three and four YOS in grades E-1 to E-3) with occupational variations; and (6) staff sergeant selection board with occupational variations. Table 5.8 provides details about the scenario features.

Officers

Table 5.9 provides results for officer scenarios. For scenarios 1, 2, and 3a, cuts focused on nontactical ops occupations could have potential adverse impact for female officers and black officers, and, to a lesser extent, Hispanic officers. The colonel SERB scenario (3b), which did not have occupational variations, shows potential adverse impact against female colonels.

Enlisted

Table 5.10 presents results for Marine Corps enlisted scenarios. As with the Army AFQT scenario, the Marine Corps AFQT scenario (4) shows potential adverse impact against all three demographic groups regardless of where the cuts are concentrated. We also analyzed this scenario using the gender-by-race/ethnicity group breakouts. For all groups except non-Hispanic white women, the trends are the same as they are for the main analysis (i.e., potential adverse impact regardless of where cuts are concentrated). For white women, adverse impact arises only when cuts are based on a 50/50 split between AFQT Categories IIIB–IV and IIIA. Specifically, the percentage of white women cut in the baseline situation (i.e., accession cuts without AFQT restrictions) would be about 4.2 percent, whereas the percentage of white women cut for a 50/50 split would be slightly higher at 4.3 percent. For the 75/25 split, the percentage of white women cut is 4.1 percent and, for the 100 percent cut from Categories IIIB–IV, the percentage of white women cut is 4.0 percent. Both of these percentages are slightly lower than the baseline percentage of 4.2. These findings show that white

---

of "General Officers and Executives, N.E.C." We therefore could not run the occupational comparison for the colonel SERB.

female accessions in the Marine Corps are not overrepresented among the lowest AFQT categories (IIIB–IV).

The first-term reenlistment control scenario (5) suggests that cuts heavily focused on nontactical ops occupations (75 percent) could result in adverse impact against all three demographic groups. Scenario 6 (staff sergeant selection board) does not show any potential adverse impact. However, a selection board like this would rely on performance and other quality indicators. An analysis of demographic variations on such indicators could reveal a different set of results.

Summary

The Marine Corps scenarios did not reveal as many potentially negative results for women and minorities as the Army scenarios. However, separations that focus more on nontactical ops occupations, particularly for officers, could have adverse impacts against women and minorities in the Marine Corps. Like the Army AFQT scenario, the Marine Corps AFQT scenario also shows potential adverse impact for all three main demographic groups. However, when the main demographic groups are broken out into gender-by-race/ethnicity groups, the AFQT scenario pattern changes for non-Hispanic white women, who could be adversely affected by AFQT restrictions but only when those restrictions are not heavily focused on the lowest AFQT categories of IIIB–IV.

Air Force

Because the Air Force has been reducing its active-duty force since 2005, we found more examples of drawdown programs than for other services. We analyzed nine Air Force scenarios. Five scenarios cover officers: (1) Force Shaping Board (FSB) for officers in O-1 through O-3 grades and three to six YOS, with occupational variations; (2) RIF of officers in O-4 through O-6 grades and six to 18 YOS, with occupational variations; (3) E-SERBs for majors with 15 or more YOS and colonels with two to four TIG, with occupational variations; (4) TERA for officers in O-3 through O-5 grades and 15–18 YOS, with

Table 5.8
Details for Marine Corps Drawdown Scenarios

Scenario Number	Program	Targeted Population	Scenario Variation(s)	Baseline Population	Cut Sizes
1	Reduction to future career force[a]	Officers at 4 YOS	• 75% of cuts from tactical, 25% from nontactical • 50% tactical, 50% nontactical • 25% tactical, 75% nontactical	All officers at 4 YOS	• 45% of targeted population (497)
2	Selective Continuation Board[a]	Majors with 20 or more YOS	See corresponding cell for scenario 1	All majors	• 10% of targeted population (58) • 20% of targeted population (116) • 25% of targeted population (145)
3a	SERB[b]	Lieutenant colonels with ≥ 4 years TIG	See corresponding cell for scenario 1	All lieutenant colonels	• 60
3b	SERB[c]	Colonels with ≥ 4 years TIG	N/A	All colonels	• 51
4*	Accessions cut and AFQT requirements tightened	Accessions (0–1 YOS)	• All cuts from Categories IIIB–IV (100%) • 75% Categories IIIB–IV, 25% Category IIIA • 50% Categories IIIB–IV, 50% Category IIIA	All accessions	• 10% accessions (2,676) • 20% accessions (5,353) • 30% accessions (8,029)

Table 5.8—Continued

Scenario Number	Program	Targeted Population	Scenario Variation(s)	Baseline Population	Cut Sizes
5	First-term reenlistment controls[a]	E-1s, E-2s, and E-3s with 3–4 YOS	See corresponding cell for scenario 1	All E-1s, E-2s, and E-3s	• 25% of targeted population (2,435) • 50% of targeted population (4,869) • 75% of targeted population (7,304)
6	Staff sergeant selection board[a]	E-6s with 15–18 YOS	See corresponding cell for scenario 1	All E-6s	• 515 • 1,030 • 1,545

SOURCES: [a] Interview notes. [b] U.S. Marine Corps (2012). [c] Sanborn (2012).

NOTE: *Scenario selected for race/ethnicity breakouts by gender.

Table 5.9
Main Results for Marine Corps Officer Scenarios

Scenario Number	Scenario Variations	Women	Non-Hispanic Black	Hispanic
1	75% tactical, 25% nontactical	6.6	4.8	5.3
	50% tactical, 50% nontactical	8.9	5.5	5.5
	25% tactical, 75% nontactical	11.3	6.3	5.7
	All officers with 4 YOS (baseline)	8.3	5.4	5.4
2	75% tactical, 25% nontactical	2.5	8.8	10.5
	50% tactical, 50% nontactical	3.7	11.7	9.2
	25% tactical, 75% nontactical	4.9	14.6	11.9
	All majors (baseline)	4.8	14.5	11.9
3a	75% tactical, 25% nontactical	1.8	4.3	4.8
	50% tactical, 50% nontactical	3.3	5.7	5.2
	25% tactical, 75% nontactical	4.8	6.8	5.3
	All lieutenant colonels (baseline)	2.5	4.8	5.2
3b	Long TIG (≥4 years)	2.9	2.9	3.1
	All colonels (baseline)	2.7	3.3	3.1

NOTES: Table values provide the percentage of cuts from each demographic group for the given scenario. Scenario 2 assumes 116 cuts. The other scenarios use cuts listed in Table 5.8.

occupational variations; and (5) VSP for officers with six or more YOS, with occupational variations. Another four scenarios cover enlisted: (6) accessions cut with tightening of AFQT requirements; (7) ERB for E-4 through E-8 grades, with occupational variations; (8) quality force review board for enlisted personnel in all grades except E-9 and with any YOS except 18–20, with occupational variations; (9) DOS Roll-back of enlisted personnel in any rank through E-8, with variations by YOS groups (three to four and five to six, seven to 12, 13–15, 20 or

Table 5.10
Main Results for Marine Corps Enlisted Scenarios

Scenario Number	Scenario Variations	Women	Non-Hispanic Black	Hispanic
4	100% cuts from Categories IIIB–IV	9.2	14.8	22.6
	75% Categories IIIB–IV, 25% Category IIIA	9.1	13.4	21.8
	50% Categories IIIB–IV, 50% Category IIIA	9.0	12.1	21.1
	All accessions (baseline)	7.8	9.0	17.2
5	75% tactical, 25% nontactical	2.5	6.1	9.4
	50% tactical, 50% nontactical	4.7	7.3	10.0
	25% tactical, 75% nontactical	7.0	8.5	10.7
	All E-1s, E-2s, and E-3s (baseline)	5.3	7.6	10.2
6	75% tactical, 25% nontactical	1.5	14.7	18.6
	50% tactical, 50% nontactical	3.0	18.0	19.3
	25% tactical, 75% nontactical	4.5	21.2	20.1
	All E-6s (baseline)	4.8	22.0	20.2

NOTES: Table values provide the percentage of cuts from each demographic group for the given scenario. Each scenario uses the middle cut size (e.g., 1,030 for scenario 6). The last column of Table 5.8 shows the cut sizes for the scenarios.

more, vs. any YOS)[10] and occupational variations. The DOS Rollback scenario has three baselines: tactical ops, nontactical ops, and any occupational group. Table 5.11 offers details about the scenario features.

---

[10] The DOS Rollback program uses reenlistment eligibility codes to target personnel within a year of their date of separation. Because we did not have the codes, we compared YOS groups that conceivably cover reenlistment points. In the Air Force, first term is usually four or six years; hence, the first YOS group is three to four and five to six. Based on guidance from a RAND Air Force personnel expert, we selected the following YOS groups for second term and beyond: second term (seven to 12), third term (13–15), and retirement-eligible (20 or more).

Table 5.11
Details for Air Force Drawdown Scenarios

Scenario Number	Program	Targeted Population	Scenario Variation(s)	Baseline Population	Cut Sizes
1	Force Shaping Board[a]	O-1s through O-3s with 3–6 YOS	• 75% of cuts from tactical, 25% from nontactical • 50% from tactical, 50% nontactical • 25% from tactical, 75% nontactical	O-1s through O-3s	• 150 • 215 • 300
2	RIF[b]	O-4s through O-6s with 6–18 YOS	See corresponding cell for scenario 1	O-4s through O-6s	• 250 • 470 • 750
3a	E-SERB[b]	Majors with ≥ 15 YOS	See corresponding cell for scenario 1	All majors	• 100 • 140 • 200
3b	E-SERB[b]	Colonels with 2–4 years TIG	See corresponding cell for scenario 1	All colonels	• 50 • 100 • 200
4	TERA[a]	O-3s through O-5s with 15–18 YOS	See corresponding cell for scenario 1	All O-3s through O-5s	• 20 • 36 • 50
5*	VSP[a]	Officers with ≥ 6 YOS	See corresponding cell for scenario 1	All officers	• 50 • 100 • 150

Table 5.11—Continued

Scenario Number	Program	Targeted Population	Scenario Variation(s)	Baseline Population	Cut Sizes
6	Accessions cut and AFQT requirements tightened	Accessions (0–1 YOS)	• All cuts from Categories IIIBIV (100%), 75% Categories IIIBIV, 25% Category IIIA, 50% Categories IIIBIV, 50% Category IIIA	All accessions	• 10% accessions (2,478) • 20% accessions (4,956) • 30% accessions (7,434)
7*	ERB[b]	E-4s through E-8s	See corresponding cell for scenario 1	E-4s through E-8s	• 700 • 1,400 • 2,100
8	Quality force review board[c]	All grades except E-9 and all YOS except 18–20	See corresponding cell for scenario 1	All grades except E-9s	• 2,000 • 3,500 • 5,000
9	DOS Rollback[d]	Enlisted personnel in YOS groups: • 3–4 and 5–6 • 7–12 • 13–15 • ≥ 20	Each YOS cluster in occupational groups: • tactical • nontactical • any occupation	All enlisted personnel in occupational groups: • tactical • nontactical • any occupation	N/A (results based on overall population proportions)

SOURCES: [a] Secretary of the Air Force Public Affairs (2014b). [b] Secretary of the Air Force Public Affairs (2014a). [c] Losey (2014b). [d] Interview notes.

NOTE: *Scenarios selected for race/ethnicity breakouts by gender.

Officers

Table 5.12 shows the main results for Air Force officer scenarios. Across scenarios, cuts that lean toward nontactical operations occupations could have adverse impacts on female officers and black officers. For the SERB scenario (3a and 3b), even a 50-percent nontactical operations cut shows potential adverse impact for black officers. This scenario also reveals longevity-based cuts matter for black officers. Unless the SERB cuts are 75-percent tactical operations occupations, the SERBs could have a negative impact on black officers because black majors and colonels are disproportionately represented among officers with longer service.

Little or no occupational trend results from analyses for Hispanic officers. The two scenarios with no adverse impact findings for Hispanic officers are scenario 1 (O-1s to O-3s) and scenario 3b (colonels). In general, cuts to officers in the late-junior to mid-career range

Table 5.12
Main Results for Air Force Officer Scenarios

Scenario Number	Scenario Variations	Women	Non-Hispanic Black	Hispanic
1	75% tactical, 25% nontactical	15.4	3.6	1.9
	50% tactical, 50% nontactical	20.8	4.6	2.2
	25% tactical, 75% nontactical	26.2	5.6	2.5
	All O-1s through O-3s (baseline)	21.5	6.0	3.4
2	75% tactical, 25% nontactical	10.4	4.0	4.6
	50% tactical, 50% nontactical	15.1	5.1	4.8
	25% tactical, 75% nontactical	19.9	6.2	4.9
	All O-4s through O-6s (baseline)	15.3	6.0	4.5
3a	75% tactical, 25% nontactical	9.4	6.6	5.9
	50% tactical, 50% nontactical	14.2	8.6	6.1
	25% tactical, 75% nontactical	19.1	10.4	6.4
	All majors (baseline)	17.2	6.9	5.1

Table 5.12—Continued

Scenario Number	Scenario Variations	Women	Non-Hispanic Black	Hispanic
3b	75% tactical, 25% nontactical	8.1	3.2	1.7
	50% tactical, 50% nontactical	11.4	4.8	1.8
	25% tactical, 75% nontactical	14.7	6.4	1.9
	All colonels (baseline)	12.1	4.6	2.7
4	75% tactical, 25% nontactical	10.0	4.4	4.7
	50% tactical, 50% nontactical	14.4	5.6	5.3
	25% tactical, 75% nontactical	19.2	6.9	5.6
	All O-3s through O-5s (baseline)	18.3	6.1	4.2
5	75% tactical, 25% nontactical	14.8	5.8	4.6
	50% tactical, 50% nontactical	19.3	7.2	5.0
	25% tactical, 75% nontactical	18.6	6.0	3.9
	All officers (baseline)	10.4	4.4	4.3

NOTES: Table values provide the percentage of cuts from each demographic group for the given scenario. Each scenario uses the middle cut size (215 for scenario 1, 470 for scenario 2, etc.). The last column of Table 5.11 shows the cut sizes for the scenarios.

(roughly O-3 to O-5) could disproportionately affect Hispanic officers, regardless of occupational focus.

Results for the Air Force officer gender-by-race/ethnicity breakouts of scenario 2 are in Table 5.13. As in the main analysis of scenario 2, women of all three race/ethnicity groups and black officers (male and female) could face adverse impact if RIFs focus heavily on nontactical ops occupations. The difference between the main analysis and the breakout analysis occurs for Hispanic officers. Like other women, Hispanic women could be negatively affected by RIFs of majors, lieutenant colonels, and colonels that focus heavily on nontactical ops occupations. However, Hispanic men could face negative results if RIFs focus more heavily on tactical ops occupations or are balanced between tactical and nontactical ops occupations.

Table 5.13
Results for Air Force Officer Scenario with Gender-by-Race/Ethnicity Breakouts

Scenario Number	Scenario Variations	NH White Women	NH Black Men	NH Black Women	Hispanic Men	Hispanic Women
2	75% tactical, 25% nontactical	8.0	2.9	0.9	4.1	0.4
	50% tactical, 50% nontactical	11.1	3.3	1.7	4.1	0.7
	25% tactical, 75% nontactical	14.3	3.6	2.5	4.0	0.9
	All O-4s through O-6s (baseline)	11.9	3.3	2.0	4.0	0.8

NOTES: Table values provide the percentage of cuts from each demographic group for the given scenario. Scenario 2 uses largest cut size in Table 5.11 (i.e., 750). NH stands for "Non-Hispanic."

Enlisted

The main results for Air Force enlisted scenarios are in Table 5.14. The AFQT scenario (6) results are similar to those for the Army and Marine Corps: All three demographic groups could be adversely affected by cuts to lower AFQT categories. Unlike the AFQT scenario, the ERB and

Table 5.14
Main Results for Air Force Enlisted Scenarios

Scenario Number	Scenario Variations	Women	Non-Hispanic Black	Hispanic
6	100% cuts from Categories IIIB–IV	23.0	34.3	1.3
	75% Categories IIIB–IV, 25% Category IIIA	22.7	31.1	1.3
	50% Categories IIIB–IV, 50% Category IIIA	22.4	28.0	1.4
	All accessions (baseline)	17.9	15.2	1.0
7	75% tactical, 25% nontactical	10.1	9.8	5.1
	50% tactical, 50% nontactical	13.3	12.2	5.5
	25% tactical, 75% nontactical	16.5	14.7	5.9
	All E-4s through E-8s (baseline)	19.4	16.8	6.3

Table 5.14—Continued

Scenario Number	Scenario Variations	Women	Non-Hispanic Black	Hispanic
8	75% tactical, 25% nontactical	10.8	9.9	4.4
	50% tactical, 50% nontactical	13.7	12.2	4.7
	25% tactical, 75% nontactical	16.5	14.5	5.1
	All enlisted except E-9s (baseline)	19.0	16.6	5.5
9	3–4/5–6 YOS, tactical	8.0	6.4	1.6
	3–4/5–6 YOS, nontactical	19.9	16.0	3.5
	3–4/5–6 YOS, any occupational group	19.7	15.9	3.5
	7–12 YOS, tactical	7.1	8.0	4.9
	7–12 YOS, nontactical	20.0	17.2	8.2
	7–12 YOS, any occupational group	19.7	17.0	8.1
	13–15 YOS, tactical	5.9	7.5	7.8
	13–15 YOS, nontactical	21.4	20.3	10.2
	13–15 YOS, any occupational group	21.0	19.9	10.2
	≥ 20 YOS, tactical	3.6	7.1	5.1
	≥ 20 YOS, nontactical	14.1	17.9	6.3
	≥ 20 YOS, any occupational group	13.7	17.4	6.2
	Any YOS, tactical (baseline)	7.6	7.6	4.2
	Any YOS, nontactical (baseline)	19.3	16.9	5.5
	Any YOS, any occupational group (baseline)	19.0	16.6	5.5

NOTES: Except for scenario 9, the table values provide the percentage of cuts from each demographic group for the given scenario, and each scenario uses the middle cut size (e.g., 4,956 for scenario 6). The last column of Table 5.11 shows the cut sizes for the scenarios. Scenario 9 is not based on a specific cut size but reflects the demographic group proportions in the targeted and baseline populations.

quality force review board scenarios (7 and 8) do not show any potential adverse impact. (Our gender-by-race/ethnicity breakout analysis for scenario 7 also showed no evidence of potential adverse impact for any racial/ethnic minority or female group.) These scenarios target a wide range of enlisted grades and YOS groups and therefore might not be narrow enough to pick up demographic group differences. In contrast to scenarios 7 and 8, scenario 9 (DOS Rollback) breaks out results by YOS groups and crosses YOS groups with occupational categories. This scenario identifies different patterns of results by demographic group. Women nearing the end of their first term could face adverse impacts, regardless of occupation. With higher YOS, the adverse impact potential lessens for women. Interestingly, cuts based on tactical operations occupations for the seven to 12 and 13–15 YOS groups do not display potential adverse impact on women, but cuts based on the nontactical operations cuts do. Since this scenario uses the proportion of women by YOS group and occupational group, the finding suggests that women in the mid-YOS groups are underrepresented in tactical ops occupational groups compared to the women in the enlisted force in general. At 20 or more YOS, adverse impact on women is not present. Overall, the DOS Rollback findings for women point to overrepresentation of women among junior enlisted ranks.

The YOS trends for black enlisted and Hispanic enlisted go in the opposite direction from trends for women. Specifically, we find more evidence of potential adverse impact against black enlisted and Hispanic enlisted in longer YOS groups than for lower YOS groups. The findings reflect underrepresentation of blacks and Hispanics at lower YOS and overrepresentation of these two groups at higher YOS.

With two exceptions, the 13–15 YOS group shows the most potential for adverse impact against women and minorities. At 15 YOS, TERA becomes available. Cuts to personnel with 13–15 YOS could therefore have the unfortunate effect of disproportionately affecting women and minorities who could be eligible for more severance compensation if allowed to remain in service another year or two.

Summary

The Air Force officer scenarios show that cuts to nontactical operations occupations could have a disproportionately negative impact on women and blacks, but not Hispanics. However, our breakout analysis suggests that Hispanic women could be adversely affected by nontactical operations occupation cuts, while Hispanic men could be adversely affected by cuts focused on tactical operations or balanced between tactical and nontactical operations occupations. Cuts based on longer service for field-grade officers (via SERBs) could particularly affect black officers. Compared to other groups, Hispanic officers may be more affected by cuts based on grade, specifically O-3 to O-5.

Unlike the officer scenarios, the enlisted scenarios did not show much of an occupational trend. The DOS Rollback scenario suggests that cuts targeting the lowest YOS group (representing first term of enlistment) could negatively affect women, whereas cuts to the highest YOS groups (representing career personnel) could negatively affect black personnel and Hispanic personnel. In addition to these experience-based results, the AFQT scenario provides a similar pattern in the Air Force as it did in the Army and Marine Corps in that accession cuts targeting lower AFQT categories could have an adverse impact against female, black, and Hispanic recruits.

Navy

Because the Navy did not plan to draw down its active-duty force in FY 2012, we use programs from the Navy's mid-2000s drawdown for scenarios. Because we have limited details on Navy programs, we examine only four scenarios. The two officer scenarios include (1) VSP for officers with six to 12 YOS, with occupational variations; and (2) SERBs for commanders (O-5s) and captains (O-6s) in tactical operations occupations, with TIG variations. The two enlisted scenarios include (1) accessions cut with tightening of AFQT requirements, and (2) ERBs for E-4 through E-5 and for E-6 through E-8, with occupational variations. Table 5.15 outlines the scenario details.

Officers
Table 5.16 shows the main results for Navy officer scenarios. The only evidence of potential adverse impact is for the VSP scenario (1), whereby cuts that are at least half nontactical ops could have an adverse impact on female officers, and cuts that are at least 75 percent nontactical ops

Table 5.15
Details for Navy Drawdown Scenarios

Scenario Number	Program	Targeted Population	Scenario Variation(s)	Baseline Population	Cut Sizes
1*	VSP[a]	Officers with 6–12 YOS	• 75% of cuts from tactical, 25% from nontactical • 50% from tactical, 50% nontactical • 25% from tactical, 75% nontactical	All officers	• 100 • 150 • 300
2a	SERB[b]	Commanders with ≥ 4 years TIG	• ≥ 4 years TIG, tactical • Any TIG, tactical	All commanders	• 30% of targeted pop. (815)
2b	SERB[b]	Captains with ≥ 4 years TIG	See corresponding cell for scenario 2a	All captains	• 30% of targeted pop. (343)
3	Accessions cut and AFQT requirements tightened	Accessions (0–1 YOS)	• All cuts from Categories IIIB–IV (100%) • 75% Categories IIIB–IV, 25% Category IIIA • 50% Categories IIIB–IV, 50% Category IIIA	All accessions	• 10% accessions (2,879) • 20% accessions (5,758) • 30% accessions (8,637)
4a*	ERB[c]	E-4s and E-5s	See corresponding cell for scenario 1	E-4s and E-5s	• 950 • 1,900 • 2,850
4b	ERB[c]	E-6s through E-8s	See corresponding cell for scenario 1	E-6s through E-8s	• 550 • 1,100 • 1,650

SOURCES: [a] Parcell (2011). [b] U.S. Navy (2011). [c] Faram (2011).

NOTE: *Scenarios selected for race/ethnicity breakouts by gender.

Table 5.16
Main Results for Navy Officer Scenarios

Scenario Number	Scenario Variations	Women	Non-Hispanic Black	Hispanic
1	75% tactical, 25% nontactical	12.3	5.2	7.1
	50% tactical, 50% nontactical	16.3	6.3	7.5
	25% tactical, 75% nontactical	20.4	7.5	8.1
	All officers (baseline)	16.2	8.1	7.9
2a	≥ 4 years TIG, tactical	2.1	3.9	5.8
	Any TIG, tactical	2.1	3.8	5.8
	All commanders (baseline)	11.6	6.5	6.1
2b	≥ 4 years TIG, tactical	1.4	3.7	3.2
	Any TIG, tactical	1.5	3.9	3.2
	All captains (baseline)	13.0	4.6	4.3

NOTES: Table values provide the percentage of cuts from each demographic group for the given scenario. Scenario 1 uses the middle cut size in Table 5.15. All cut sizes are listed in the last column of Table 5.15.

could have an adverse impact on Hispanic officers. The SERB scenario (2) focused on tactical operations occupations only. Removing the TIG restriction did not change results much, if at all.

Results for our gender-by-race/ethnicity breakouts of scenario 1 are in Table 5.17. Unlike the main analysis of scenario 1, the breakout analysis shows that for all groups except Hispanic men, heavy nontactical cuts (75 percent) could result in adverse impact. For Hispanic men, the trend is in the opposite direction, whereby heavier tactical ops cuts could negatively affect Hispanic men. These trends mirror those found for the Air Force officer RIF scenario (2). However, the number of individuals cut within each of these breakout groups is small (i.e., 35 or fewer). Therefore, the addition or subtraction of one person could shift the percentages and trends. We recommend caution in interpreting these findings, particularly for non-Hispanic black women and Hispanic women.

Table 5.17
Results for Navy Officer Scenario with Gender-by-Race/Ethnicity Breakouts

Scenario Number	Scenario Variations	NH White Women	NH Black Men	NH Black Women	Hispanic Men	Hispanic Women
1	75% tactical, 25% nontactical	7.8	5.7	1.1	6.7	1.1
	50% tactical, 50% nontactical	9.7	6.1	1.8	6.7	1.4
	25% tactical, 75% nontactical	11.6	6.4	2.5	6.6	1.8
	All officers (baseline)	10.5	6.2	2.1	6.6	1.6

NOTES: Table values provide the percentage of cuts from each demographic group for the given scenario. Scenario 1 uses largest cut sizes in Table 5.15 (i.e., 300). NH stands for "Non-Hispanic."

Enlisted

Table 5.18 presents results for the Navy enlisted scenarios. The Navy AFQT scenario (3) follows the same pattern as the AFQT scenarios for the other services by showing potential adverse impact for all three demographic groups. The ERB scenario (4) did not show any adverse impact results. However, cuts that lean heavily toward nontactical ops result in cuts closer to baseline than cuts that lean heavily toward tactical ops, suggesting some occupational impacts are possible.

We ran gender-by-race/ethnicity breakout analyses on scenario 4 for Navy enlisted. As shown in Table 5.19, the findings from the breakout analysis have the same trends as for the main analyses (i.e., no adverse impact) except for Hispanic men. Regardless of the type of occupational cut, cuts to enlisted personnel in E-4 through E-8 grades in the Navy could negatively affect Hispanic men.

Summary

The Navy scenarios offer little evidence of potential adverse impact against women and minorities. Exceptions include scenario 1 (VSP for officers with six to 12 YOS) and scenario 3 (AFQT scenario for enlisted accessions). Occupational variations in scenario 4 did not offer evidence of any adverse impact, but cuts that are heavily nontactical are closer to baseline cuts than cuts that are heavily tactical, suggesting

Table 5.18
Main Results for Navy Enlisted Scenarios

Scenario Number	Scenario Variations	Female	Non-Hispanic Black	Hispanic
3	100% cuts from Categories IIIB–IV	31.9	27.0	18.4
	75% Categories IIIB–IV, 25% Category IIIA	30.4	25.0	19.7
	50% Categories IIIB–IV, 50% Category IIIA	28.8	23.0	21.0
	All accessions (baseline)	21.8	13.9	19.5
4a	75% tactical, 25% nontactical	14.5	15.3	20.4
	50% tactical, 50% nontactical	15.5	15.7	20.4
	25% tactical, 75% nontactical	16.5	16.1	20.5
	All E-4s and E-5s (baseline)	17.1	16.3	20.6
4b	75% tactical, 25% nontactical	6.9	14.8	13.4
	50% tactical, 50% nontactical	8.6	16.3	13.5
	25% tactical, 75% nontactical	10.3	17.8	13.5
	All E-6s through E-8s (baseline)	11.4	18.7	13.6

NOTE: Table values provide the percentage of cuts from each demographic group for the given scenario. Each scenario uses the middle cut size (e.g., 5,758 for scenario 3) listed in the last column of Table 5.15.

the cuts to nontactical occupations would be more prone to result in adverse impact. Furthermore, our breakout analysis of scenario 4 suggests that Hispanic men, regardless of occupational variation of cuts, could be adversely affected by cuts to Navy enlisted personnel in the E-4 through E-8 grades.

Table 5.19
Results for Navy Enlisted Scenario with Gender-by-Race/Ethnicity Breakouts

Scenario Number	Scenario Variations	NH White Women	NH Black Men	NH Black Women	Hispanic Men	Hispanic Women
4a	75% tactical, 25% nontactical	4.8	11.4	3.8	11.7	2.2
	50% tactical, 50% nontactical	5.1	11.6	3.9	11.5	2.4
	25% tactical, 75% nontactical	5.4	11.9	4.1	11.3	2.5
	All E-4s and E-5s (baseline)	5.6	12.1	4.1	11.1	2.6
4b	75% tactical, 25% nontactical	3.1	12.5	2.1	9.6	0.8
	50% tactical, 50% nontactical	3.6	13.3	2.8	9.4	1.0
	25% tactical, 75% nontactical	4.0	14.1	3.5	9.3	1.2
	All E-6s through E-8s (baseline)	4.3	14.6	4.0	9.2	1.3

NOTES: Table values provide the percentage of cuts from each demographic group for the given scenario. Scenario 4 uses largest cut sizes in Table 5.15 (e.g., 1,650 for scenario 4b). NH stands for "Non-Hispanic."

Policy Implications of Results

Implications of Results from Scenario Analysis

Three themes from our analyses have policy implications for drawdowns of active-duty military personnel. The first theme involves occupations. Across several scenarios, cuts that are focused heavily on personnel in nontactical operations occupations (e.g., logistics, nursing, maintenance) could have an adverse impact on women and racial/ethnic minorities. Although the Navy scenarios do not show potential adverse impact from occupational cuts, heavy nontactical operations cuts tend to be closer to baseline cuts, meaning that nontactical operations cuts could be more negative for women and minorities than cuts heavily focused on tactical operations occupations. In general, if the services place a heavy emphasis on cutting nontactical operations occupations—all else being equal—blacks and women (and, to a lesser extent, Hispanics) could be negatively affected.

The second theme involves the role of service length, either in terms of YOS or TIG. For the Army and Air Force, reductions involving longer service could negatively affect retention of black personnel. For example, the SERB scenarios in both services show potential adverse impact for black officers. The Army retention control point scenario and the Air Force DOS Rollback scenario reveal similar trends, but for black enlisted personnel. Conversely, cuts to junior personnel could have an adverse impact on women. For example, the Air Force DOS Rollback cuts to first-term enlisted (i.e., those with three to four or five to six YOS) show adverse impact on women across all occupational variations in the scenario. However, cuts to higher YOS groups are not uniformly negative for women, with the 20+ YOS group showing no evidence of adverse impact for female enlisted personnel in the Air Force. This theme is perhaps the most challenging to address with policy because the adverse impact trends go in opposite directions for women and blacks: Cutting junior personnel could have an adverse impact on women but not blacks, and vice versa.

The final theme involves the AFQT accession cut scenario. For all four services, accession cuts that involve a limit on recruiting individuals with low AFQT scores could result in adverse impact on female, black, and Hispanic recruits. Although many organizations become more selective when they need to hire fewer employees, the choice to tighten enlisted entry standards has the potential effect of reducing female and minority representation among accessions, which has implications for demographic diversity in the junior enlisted ranks.

Discussions with Force Management Experts and Diversity Experts

To identify policy options to address potentially negative impacts of drawdown strategies on women and racial/ethnic minorities, we met with some of the experts whom we interviewed about recent or current drawdown goals, strategies, policies, and practices. In fall 2014, we met with two experts from each service. The experts have leadership roles in force management policy and execution, or in diversity-related efforts. These experts reside in manpower and personnel directorates in the services. Meetings were held by phone or in person and lasted anywhere from 30 to 60 minutes. We asked experts the following types of questions:

- Does your service currently look at potential demographic impacts of drawdown strategies? If so, what has your service examined?
- Within the boundaries of current law and policy, could your service modify its drawdown strategies to address differential impacts on demographic groups?
- If not, what would need to change in law or policy to make that possible?
- What can your service do within current legal boundaries to address potential adverse effects of drawdowns on certain demographic groups?

In all of our discussions, experts stated that the services could do little, if anything, to address demographic diversity in drawdown decisions because of legal barriers. Specifically, the services cannot use demographic information to decide who stays and who goes. Experts emphasized merit and mission requirements as primary drivers of drawdown decisions.

When we asked where policy could change to limit potentially negative effects of drawdowns on demographic diversity, the discussions tended toward the usual areas for diversity policy: recruiting, career assignment, and career flexibility. Examples of policy options mentioned by experts include

- Focus more outreach and recruiting efforts on women and minorities with science, technology, engineering, and math (STEM) degrees (Air Force).
- Encourage more women and minorities to enter operational career fields (Army).
- Review DOPMA provisions to determine whether there are ways to add more flexibility to promotion timing for officers (Air Force and Army).

- Provide more "on-off ramps" to help retain personnel in the Total Force (Navy).

The experts also highlighted the Career Intermission Pilot Program as an example of a program designed to provide more flexible career options in order to better retain talented personnel (but with the hope that women in particular will retain at higher levels). The program allows each service up to 20 enlisted personnel and 20 officers to take a sabbatical of up to three years to pursue education, family, and other life goals. The Navy is the first service to implement the program, which it did after its introduction in the FY 2009 NDAA. In FY 2014, the other three services joined the Navy in implementing the program. The efficacy of the program is still unknown and, in the case of the Marine Corps, only three Marines have signed up for the program as of fall 2014.

The experts noted that the "on-off ramp" concept is a major focus now because of the drawdown. Specifically, the services are looking for ways to move talented personnel from active to reserve status during the drawdown in the hopes they can bring some of those individuals back to active duty once the drawdown subsides. At this point, however, it is not clear what impact these on-off (transition) programs would have on lessening any adverse impact of the drawdown on minorities or women in the active-duty force.

Conclusion

The scenario findings suggest three policy-relevant themes. First, cuts to nontactical operations occupations can have an adverse impact on female personnel and black personnel, and, in some cases, Hispanic personnel (mainly Hispanic women, not Hispanic men). Second, cuts to personnel with longer service could adversely affect black personnel, but cuts to personnel with shorter service can adversely affect women. Third, tightening AFQT standards as part of an accession cut strategy could result in adverse impact on female, black, and Hispanic recruits, although non-Hispanic white women may not be as adversely affected as members of racial/ethnic minority groups.

The services are limited in how they can use demographic information in a drawdown context. The legal fallout of the lawsuits from the 1990s separation boards, and the trend in affirmative action cases in the United States more generally, make it unlikely that the services will pursue demographically driven decisions during a force reduction. That said, at least one force management expert we interviewed stated that his service is monitoring demographic outcomes from the drawdown. Based on this and other comments from force management experts, our recommendation would be for the services to conduct predecisional analyses to anticipate potential adverse results on demographic diversity from drawdown decisions. We discuss this and a related recommendation in the next chapter.

CHAPTER SIX

Conclusions and Recommendations

The study presented in this report sought to address how future military drawdowns could affect demographic diversity of the DoD workforce. To address this question, we began by looking to the past by examining drawdown goals, strategies, and workforce changes during and after the 1990s and mid-2000s drawdowns. Based on our findings from this historical analysis, and our review of goals and strategies for recent and upcoming force reductions, we developed and analyzed notional drawdown scenarios to illuminate possible adverse impacts on female, black, and Hispanic active-duty personnel. Below, we summarize findings from our analyses of historical drawdowns and future drawdown scenarios and offer recommendations for DoD policy.

1990s Drawdown

Active-Duty Force

The last major reductions to affect all four DoD services occurred after the Cold War, in the late 1980s through most of the 1990s. The services' main goal during that period was to protect its career force by not involuntarily separating many of the men and women who gave years of service. Consequently, the services relied on cuts to accessions, early retirements for those who were already retirement eligible, and voluntary separation measures. What resulted was a force that was more senior in experience, balanced more toward officers than enlisted personnel, and (at least in the officer corps) more heavily represented in nontactical operations occupations.

What also occurred in the 1990s was an increasingly demograph-ically diverse force. With a few exceptions, female and racial/ethnic minority representation in the active-duty force increased from FY 1990 to 2001. In general, female representation and Hispanic representation increased from higher representation among accessions, whereas black representation increased from higher retention rates relative to other racial/ethnic groups. Unlike black retention, female retention generally remained lower than male retention. Our adjustments to female reten-tion rates to account for workforce characteristics like grade, education level, and YOS do not fully explain the gender retention gap, which persists to this day.

Because the services used a variety of drawdown strategies and tools during the drawdown, we could not pinpoint the collective effects of various drawdown policies on demographic diversity in the 1990s. Only one type of drawdown policy ties directly to the 1990s draw-down: Army and Air Force policy to ensure equal opportunity goals are met by officer separation boards. However, the strategy chosen to address this policy put both services in legal situations that they do not wish to repeat. The drawdown policy to "keep faith" with the career force likely affected demographic diversity as the policy drove most of the drawdown strategies used by the services, notably large-scale accession cuts and use of voluntary separation incentives like VSI/SSB, which have demonstrated demographic impacts.

The Reserve Component and Civilian Workforce

The reserve reductions were not as severe as they were for the active component. For example, the Army National Guard shrank by about 20 percent and the Army Reserves by about 33 percent from their peaks in 1991 to 1999. In comparison, the Army's active force reduced by about 39 percent from its peak in 1987 to 1999. The relatively smaller reserve reductions were partly due to the National Guard pushing back on larger cuts from Congress and DoD's intent to shift certain mis-sions and elements of force structure from the active component to the reserve component. The Army's support units experienced the brunt of the reserve force drawdown.

Although the reserve component drawdown was smaller than the active component drawdown, the services used a variety of strategies to cut both forces. A popular strategy to shrink the force sizes in both components involved restrictions in the size of accessions. Another similarity between the two components is that both experienced increases in demographic diversity in the 1990s. Therefore, neither active component nor reserve component reductions appear to have negatively affected demographic diversity on the whole.

Compared to the reserve component drawdown of the 1990s, the DoD civilian drawdown was larger; the civilian workforce downsized by 36 percent between FY 1989 and 1999. Civilian reductions were primarily driven by the political imperative of lower military spending during an era of relative peace as well as by Congress' desire that civilian employment come down at a rate equivalent to the overall military drawdown. Cuts were distributed approximately equally across the services and were primarily achieved through hiring freezes and voluntary separation measures first, and reductions in force as a last resort. The impact of the civilian drawdown was felt most by clerical and blue-collar workers and junior employees. Earlier reports on the civilian reductions suggest that black and female personnel may have been disproportionately affected, although these earlier reports do not paint a clear picture of the overall demographic outcomes of the civilian drawdown.

Drawdowns in the 2000s and 2010s

Mid-2000s Reductions of the Active-Duty Forces in the Navy and Air Force

In the mid-2000s, the Navy and Air Force shrank again as the Army and Marine Corps grew in size to conduct operations in Iraq and Afghanistan. The Navy reductions started first, around FY 2003, and the Air Force reductions began in late FY 2004. Although both services had many of the same goals as they did in the 1990s, they had more concerns about balancing the force and retaining talent. Their strategies involved fewer accession cuts than in the 1990s but more voluntary and involuntary separation programs targeting personnel in overage skill areas and with less desirable records (e.g., lower performance evaluations) than similarly situated peers.

By FY 2011, the Navy active-duty force was a more officer-heavy force, as it was after the 1990s reductions. Unlike the 1990s, the Navy officer corps was more heavily weighted toward tactical operations occupations, and the experience of the force was not much higher by FY 2011. The Air Force, by comparison, balanced its enlisted and officer reductions so that the enlisted-to-officer ratio did not change much. The Air Force enlisted force was more junior and, like the Navy's, its officer force leaned more toward tactical operations occupations.

The Air Force of the late 2000s also differed from the Air Force of the late 1990s/early 2000s in another way: demographic diversity. The Air Force did not increase the demographic diversity of its enlisted force between FY 2001 and 2011. The Navy increased its demographic diversity for the most part, the exception being a drop in enlisted black representation between FY 2001 and 2011. Because the Air Force enlisted force lost demographic diversity, we examined the enlisted force further by focusing on gender diversity as an example. As we found in our analysis of gender retention trends for Army officers during and after the 1990s drawdown, the gender retention gap played a role in lower female representation in the Air Force enlisted force in the 2000s. Adjusting the female retention rates, as before, did not fully explain the gender retention gap. Unlike the Army officer analysis, the Air Force enlisted analysis also revealed a drop in female shares of accessions. The accession losses may be partly attributable to tightening standards, as seen by higher female representation among lower AFQT accessions compared with female representation in the enlisted force.

As with the 1990s drawdown, the mid-2000s drawdown in the Navy and Air Force offered few opportunities to identify the degree of impact that drawdown policies had on demographic diversity. We speculate that the Air Force's large cut to enlisted accessions near the beginning of the drawdown (FY 2005) and its tightening of entry standards may have factored into lower female shares of accessions. However, without details on all major drawdown programs and practices—who was targeted and who separated/was denied entry into the service due to the programs/practices—we cannot state that any one program or practice directly affected the representation of women or racial/ethnic minorities in the services.

Recent Reductions in the Reserve Component and Civilian Workforce

Although the size of recent reserve reductions is not yet clear, the experts we interviewed do not expect the reductions to be as large as those in the 1990s. Except for the Navy Reserve, which already cut its personnel in earlier years, the other service reserve components are trying to attract midgrade personnel leaving active duty and believe they can address any reserve reductions through a combination of natural attrition, cuts to accessions, transfers to the individual ready reserve, and, as needed, administrative (involuntary) separations. Based on demographic data for the Army National Guard, minority and female representation has grown anywhere from 0.4 to 1.9 percentage points between FY 2010 and July 2014. However, within that same time period, black and female enlisted guard members have been disproportionately affected by administrative separations.

Although the recent civilian drawdown is not expected to be as large as the one in the 1990s, the discretionary spending cuts associated with the 2011 Budget Control Act will probably result in substantial and permanent reductions in civilian employment within DoD. The Army in particular is expected to have a larger share of civilian reductions than the Air Force and Navy. The services expect to utilize similar drawdown tools as in the 1990s, with the hope that voluntary separation incentives, early retirements, and hiring freezes will achieve much of the civilian reductions. In particular, the services do not wish to repeat the furloughs of 2013. The demographic impact of recent civilian drawdowns is not yet clear, although the services are concerned about losing highly skilled younger employees, and the Air Force is also concerned about separations of Hispanics, women, and persons with disabilities. Our analysis of separation data for Army civilians between FY 2011 and the second quarter of FY 2014 shows that black men and women faced a disproportionate number of administrative separations.

Scenarios for Future Active-Duty Drawdowns

Because our review of historical drawdowns of the active-duty force did not offer the level of detail needed to identify effects of specific drawdown plans on demographic diversity, we turned toward understanding recent and upcoming drawdown plans to identify potential future impacts of drawdowns on demographic diversity. To develop scenarios, we spoke to force management experts in the services and DoD and reviewed drawdown authorities and tools available to the services. As with the historical drawdown review, the services did not supply enough details about specific drawdown plans to create detailed scenarios. Moreover, the DMDC personnel data that we used lack performance evaluations, disciplinary records, and other personnel information needed for analysis of drawdown scenarios based on "quality" factors. Despite these limitations, we developed notional scenarios based on recently completed or upcoming drawdown programs as identified through our interviews with experts or publicly available information, such as news stories. We added some occupational and experience-based comparisons to most scenarios to examine what could happen were drawdown cuts focused more heavily on some occupational or experience categories than others. We compared the scenarios' target populations (e.g., colonels with four or more years TIG) to baseline populations (e.g., all colonels) to determine the potential for adverse impact on women, blacks, and Hispanics. We also explored how interactions of race/ethnicity and gender could affect scenario results by analyzing a select number of scenarios for the following demographic groups: non-Hispanic white women, non-Hispanic black women, Hispanic women, non-Hispanic black men, and Hispanic men. Cases where cuts to the target population exceed cuts to the baseline population provide evidence of potential adverse impact.

Across services, we identify three policy themes from the scenario findings. First, cuts to nontactical operations occupations could have adverse impact on women and blacks, and, to some extent, Hispanics (although our breakout analyses suggest that Hispanic men may be adversely affected by cuts to tactical operations occupations instead). Second, cuts to personnel with longer service (YOS or TIG) could negatively affect black personnel, but the opposite holds for women. Third, tightening AFQT standards during an accession cut could have negative consequences for demographic diversity of enlisted accessions, but to a lesser extent for diversity based on non-Hispanic white women. The first two themes suggest that force-shaping policies based on occupational area or length of service could have a negative impact on women and minorities depending on which occupational or experience groups are targeted for separations. The third theme suggests that the common accession policy to tighten accession standards while cutting accessions during a drawdown could have an adverse impact on female and minority recruits.

Recommendations

Because of legal limitations on using demographic information in force management decisions, we do not recommend specific changes to force management policy that would require the services to make drawdown decisions based on a person's gender, race, ethnicity, or other protected status. However, we make two related recommendations regarding how force management can consider demographic implications of drawdown decisions.

OSD(P&R) Directs the Services to Conduct Predecisional Analyses

We recommend that OSD(P&R) direct the services to conduct adverse impact analyses *prior to* making drawdown decisions, not after decisions are made. Adverse impact analyses in employment contexts are commonly associated with hiring (organizational entry) decisions. However, per the *Uniform Guidelines on Employee Selection Procedures* (EEOC, 1978), employment decisions at other career junctures (e.g., promotions, separations) could benefit from adverse impact analyses. If an employment procedure is demonstrated to have adverse impact on the job opportunities of a legally protected class (e.g., racial minority), these guidelines stipulate that the employer establish the "business necessity" of continued use of that measure. To establish business necessity, the selection procedure should be validated. This requires validating the tests or measures used in the procedure and demonstrating that attempts were made to find ways to reduce adverse impact (Zedeck, 2009).

Although the EEOC requirements for adverse impact analysis do not apply to employment decisions regarding military personnel, DoD is not precluded from using the EEOC guidance for such purposes. Moreover, validating employment decisions according to the *Uniform Guidelines* could serve DoD well if the legal context were to change such that DoD would be required to comply with EEOC regulations for employing military personnel. We therefore identify the following types of questions that the services would need to address when analyzing adverse impact of drawdown procedures:

- Is there evidence that the measures of a person's merit are empirically valid? How about the validity of the weights placed on the different measures in making drawdown selection decisions?
- Are there other measures (or combinations thereof) that are valid but have less adverse impact?
- Can the occupations, experience levels, or other factors used to target personnel be directly tied to mission requirements? Are those "requirements" valid?
- What policies or laws would have to change to address adverse impact?

In drawdowns, the services aim to retain as much talent (i.e., high-performing personnel) as possible. However, there are many ways that the services try to measure talent or merit, including performance evaluations, disciplinary records, and tests to meet annual standards (e.g., physical fitness tests).[1] The services use combinations of these types of

---

[1] Although commonly used by the services to assess military personnel's physical ability to serve, physical fitness tests (PFTs) are not measures of job performance. Job performance measures reflect the ability to perform job tasks. PFTs are an assessment of overall physical fitness, not one's ability to perform specific physical tasks on a job. In 1999, the Congressional Commission on Military Training and Gender-Related Issues recommended that the services educate personnel on the difference between physical fitness assessment and job performance standards (Congressional Commission on Military Training and Gender-Related Issues, *Final Report: Findings and Recommendations*, Volume 1, July 1999).

measures and often board processes to make decisions on whom to select for separation. The measures used in the drawdown decisions should be valid indicators of the various objectives that the services wish to meet during a drawdown (e.g., balanced force structure, maintaining personnel in critical skill areas, fair and equitable treatment of all personnel), as should the weights placed on the different measures in the drawdown decision. Validity evidence can come from many sources, but a common approach is to establish criterion-related validity. That is, how well does the measure predict important organizational outcome criteria, like retention, promotion, and performance? If, for example, a person's prior disciplinary issues do not predict his or her future performance, past disciplinary infractions may not be useful in drawdown decisions. A related issue is the weight assigned to the different measures used in the decisionmaking process. For example, how much weight does a separation board place on a past disciplinary infraction vice a recent performance evaluation? The services may already know the answers to these questions but still need to document the evidence if there is concern about adverse impact.[2] Even after validity evidence is gathered, a drawdown strategy (e.g., SERB) may still be used even if it has adverse impact against individuals from a certain group (e.g., black colonels) because the measures and decision points that are part of the strategy are shown to meet valid organizational objectives.

A second prong of adverse impact analysis is determining whether alternative (but also valid) measures or combinations of measures could be used to reduce adverse impact. The services can use evidence from empirical studies about measures that show less adverse impact on women and minorities in other employment contexts (e.g., hiring) and

[2] A finding of adverse impact does not necessarily mean that the methods and measures used to make employment decisions are not valid. In civilian contexts, an organization may continue to use methods and measures that result in adverse impact if those methods and measures are tied to important organizational criteria (i.e., are a business necessity). If the services choose to use drawdown methods and measures that are shown to be valid and necessary but have an adverse impact on members of certain demographic groups, they should be prepared to handle the attention they may receive from various stakeholder communities. The services have experience in handling such attention in other employment decision contexts (e.g., adverse impact of the ASVAB).

data from prior drawdowns to identify alternative measures or combinations of measures.

Another set of assumptions to validate in a drawdown context is how mission requirements that affect drawdown decisions are determined. For example, a new weapon system may require expertise in cyber, thus increasing the criticality of cyber skills and leading to a decision to protect cyber career fields from a drawdown. Even if the service cannot validate the requirement for the weapon system (i.e., requirement is outside the service's control), the logic behind how the requirement ties to specific drawdown decisions should be documented so OSD(P&R) knows the assumptions behind specific types of drawdown decisions.

The final question to address is how and which policies and laws would need to change to address adverse impact in a drawdown context. These policies and laws may not directly tie to drawdowns but could have implications for the demographic outcomes of future drawdowns. For example, Air Force experts said they were looking at whether changing the way officers are grouped into competitive categories for promotion could help retain more female officers. Since competitive categories factor into drawdown authorities like SERBs, changes to competitive categories could have implications for how drawdown decisions affect demographic diversity in the officer corps. An examination of SERB scenarios with different competitive category configurations could help identify possible demographic outcomes of SERBs, were promotion policy to change.

To assist the services in conducting these analyses, MPP within OSD(P&R) should develop policy to guide the services' efforts. ODMEO should assist in the development of this policy guidance because of its expertise in (civilian) equal employment opportunity, where adverse impact analysis is common practice. The guidance should have at least three components. First, the guidance should outline the types of questions that the analyses should address, such as the questions described above. Second, the guidance should offer a general approach to analysis but need not prescribe that approach (i.e., allow the services some leeway). The guidance could use our approach as a general template for how the services can think about their own analyses. Specifically, the services can review their historical draw-downs to identify demographic trends and policy themes. The services can then create drawdown scenarios with variations based on occupa-tion, experience, or other factors that go into drawdown decisions. The services should be encouraged to develop more-detailed scenarios and use analytical modeling as appropriate. Finally, the guidance should state that services briefly describe which measures and data elements they used in their analyses. The overall goal of the OSD policy is to ensure that the services "do their homework" by conducting analyses and providing appropriate documentation. If the services do not suf-ficiently justify their strategies, OSD would have the option of "non-concurrence" with the proposed strategies. A nonconcur determination would require the services to either change strategies or provide OSD with the appropriate documentation to support their current strategies.

ODMEO Validates Services' Results, and OSD Directs DMDC to Acquire More Data

Given its expertise in adverse impact and related concepts, ODMEO should be responsible for validating the services' predecisional ana-lytic results. Because validation can be time-consuming, MPP and ODMEO can state conditions under which validation checks will not be conducted (e.g., for drawdown decisions that would affect less than 5 percent of a targeted population). At a minimum, ODMEO would make sure the services answered the appropriate questions stated in the policy guidance. Instead of checking all analyses and results, ODMEO can spot-check some analyses using service-provided or DMDC data. To use DMDC data, DMDC would need to acquire more detailed data that measure factors used in drawdown decisions. As we previ-ously noted for our own analysis, DMDC personnel data lack per-formance evaluation, disciplinary records, and other personnel infor-mation needed for analysis of drawdown scenarios based on "quality" factors. OSD(P&R) should direct DMDC to acquire these data from the services (and direct the services to provide the data). Moreover, the DMDC records would need to have the same level of detail as found in the services' data files. For example, if the services target personnel based on detailed reenlistment codes, DMDC would need to make sure their reenlistment codes are at the same level of specificity.

Final Remarks

Examining how force reductions affect the demographic composition of the workforce is not a simple endeavor because the size and scope of drawdowns evolve, and the services use a variety of drawdown tools, many at the same time. However, our investigation of the drawdowns in the 1990s and mid-2000s show that drawdowns do not spell disaster for the demographic profile of the military force. In fact, demographic diversity generally increased for much of the 1990s. Instead of large-scale shifts in the demographic profile, drawdowns may affect demographics of subpopulations, like colonels with longer TIG. It is at the level of particular drawdown decisions where adverse impact issues may reside.

After the legal fallout from the 1990s drawdown, the services are not inclined to have demographic diversity goals for drawdowns. However, the law does not restrict demographic-trends analyses to determine how drawdown decisions could affect different demographic groups. By doing predecisional analyses on potential demographic impacts, the services may identify interventions for earlier in the career life cycle, such as at initial career assignment, where they could have a greater long-term impact on demographic diversity. Without any consideration of demographic diversity during a drawdown, DoD runs the risk of inadvertently undermining diversity goals, including the goal of having a military force that reflects the nation it serves.

Reserve Component Drawdowns

The purpose of this appendix is to compare the consequences of post–Cold War and recent personnel drawdowns for DoD's reserve components (RCs) in order to better understand factors affecting demographic diversity during periods of military downsizing that may be somewhat different for reserve forces than they are for active forces. Relying on interviews with reserve personnel officials in the Army, Air Force, Navy, and Marine Corps, limited aggregated reserve personnel data, and open source and DoD documentation on reserve personnel policies and effects, this appendix provides a largely qualitative and descriptive account of DoD's attempt to shape its reserve component workforce over the last 25 years.

The appendix has three main sections. The first focuses on the 1990s drawdown, and the second focuses on the current drawdown. Each of these sections addresses the following four personnel management topics:

1. the principal drivers of change in the size and shape of DoD's reserve forces
2. the nature, scope, and timing of personnel changes within service RCs
3. measures taken by DoD to effect desired personnel changes within the RCs
4. actual changes in the composition of RC workforces during the drawdown period.

The appendix concludes with a comparison of RC responses to military personnel drawdowns—and their impact—across time and service.

Post–Cold War Era

This section addresses the politics, policies, and outcomes associated with managing reserve force personnel during the late 1980s and 1990s drawdown.

Army National Guard and Reserve
Active-Reserve Force Mix

At the start of the post–Cold War era, military and political leaders and defense policy analysts debated the appropriate mix of active and reserve forces within DoD generally and within the Army in particular. To many in Congress, the reduced global threat meant that the United States could rely more on reserve forces for its security. To DoD's leadership, however, retaining reserves at higher-than-requested levels would increase costs and lead to an unbalanced force structure. Moreover, the Army's lesson from the 1990–1991 Persian Gulf War was that depending on reserve combat brigades, which required extensive predeployment training, in overseas crisis situations was a mistake. Accordingly, the Base Force strategy, announced in 1990, prioritized units that could respond quickly to major regional contingencies, most of which belonged to the active component (GAO, 1992). In support of the congressional case, GAO doubted that retaining reserves above planned levels would result in reserve units without missions, noting that the Army had deployed almost all of some types of support forces, many of which were in the reserves, in the Gulf War, and at least 90,000 support positions within the service remained unfilled. Additionally, GAO touted the long-term savings that could be achieved if some missions were shifted from active forces to the reserves.[1]

---

[1]   According to budgetary data for FY 1993, the average cost of pay, allowances, and benefits for a reservist was $8,300, compared to $39,000 for an active-duty soldier (GAO, 1992).

Reserve Force Planning

In the midst of the force mix debate, the Army proceeded with plans for accelerated force reduction options for the active and reserve components. The "Quiet Study" proposal, produced by the Army headquarters' program analysis division, made ten specific recommendations for cuts. The recommendations targeted forces that would not be decisive in a global war, those with aging equipment, and those units whose growth was outpacing the growth of the Soviet threat, such as armor forces. Reserve forces figured prominently in several other courses of action proposed between 1989 and 1994. These were prompted by the results of the first Gulf War as well as the Base Force analysis and the 1993 Bottom-Up Review. Each proposal entailed an RC force reduction. By 1994, the Army was proposing end-strengths of approximately 397,000 and 260,000 for the Army National Guard (ARNG) and U.S. Army Reserve (USAR), respectively, down from 457,000 and 319,000 in 1989 (see Figure A.1).

Figure A.1
Army Force Structure and End-Strength Proposals, FY 1989–1994

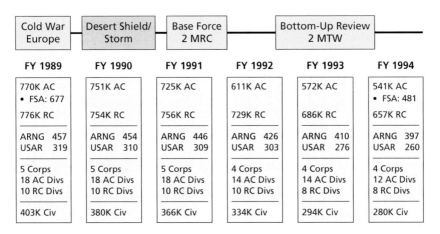

SOURCE: Unpublished internal ARNG document, 2014.
NOTES: Force size estimates (e.g., 770K) represented in thousands of people. The abbreviations stand for the following terms: Civ = Army civilians, Divs = Army divisions, FSA = force structure allowance, MRC = major regional conflict, and MTW = major theater war.
RAND RR1008-A.1

Extent of Reserve Reductions

Despite the Army's planning efforts, Congress did not approve all of the Army's proposed reserve force reductions. For example, Congress authorized a reduction of only about 43,000 reserve positions compared with the 82,000 proposed in DoD's 1992 budget request. Similarly, in 1993, Congress authorized about 31,000 positions cut, compared with the 92,000-position reduction that DoD had requested (GAO, 1992). In response to the 1997 Quadrennial Defense Review (QDR), the Army announced plans to cut the USAR by 7,000 personnel and the ARNG by 38,000 personnel. However, ARNG officials argued that they had not been included in the QDR's process and would not feel bound by its conclusions (GAO, 1998). In the final analysis, the Army's reserve component was reduced significantly at the end of the Cold War, but not to the same extent as the active component. Between its high points in 1991 and 1999, the ARNG declined from approximately 446,000 to 357,000, or 20 percent. For its part, the USAR lost about a third of its forces, shrinking from about 310,000 to 207,000. By contrast, the active component of the Army declined by 39 percent, from approximately 770,000 at its peak in 1987 to 474,000 in 1999 (see Figure A.2).

Drawdown's Impact on the Army Reserve Components' Workforce

According to GAO, the impact of the post–Cold War drawdown fell most heavily on Army support units (GAO, 1995). Given that many of the Army's support units are in the USAR, targeting support forces may have had a disproportionate effect on women and racial/ethnic minorities because they tend to be concentrated in the support branches. However, evidence suggests that demographic diversity increased in the Army National Guard and Reserves between FY 1990 and 2001. Table A.1 shows the percentages of women and minorities in the Army National Guard and Army Reserve in 1990, 1995, 2000, and 2001. As shown in the last row of the table, the percentages of women and minorities were higher in 2001 than in 1990, upwards of 12.7 percentage points in the case of enlisted minorities in the Army Reserve. Thus, any occupational impacts that the drawdown may have had in

Figure A.2
Army Personnel Strength by Component, FY 1979–1999

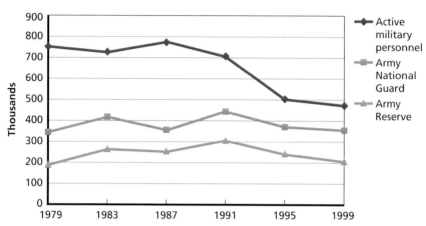

SOURCE: Data compiled from Table 1 in Brinkerhoff (2000, p. 10).
NOTE: Army National Guard and Army Reserve data include selected reservists only;
data do not include personnel in the Individual Ready Reserve, Individual National
Guard, or Retired Reserve.
RAND RR1008-A.2

the 1990s did not translate into an overall loss in female and minority
representation in the Army National Guard and Reserve.[2]

Air National Guard and Air Reserve

The post–Cold War military drawdown had little effect on the size
of the Air National Guard and Air Reserve. Although their numbers
decreased by a few thousand in the 1990s, their sizes relative to the
active component increased. Although this was due in part to congressional resistance to significant reserve force reductions, the Air Force
also implemented a policy of making greater use of reserve forces as a
result of the 1993 Bottom-Up Review and the high operational tempo
it was experiencing in the Middle East and the Balkans in the mid-
1990s. During this period, the Air Force transferred the F-15s required

---

[2]   According to Army National Guard experts, the black enlisted Guard and, to a lesser
extent, the Hispanic enlisted Guard decreased between FY 1989 and 1995. However, female
representation in the enlisted Guard increased. Also, blacks, Hispanics, and females also saw
representation gains in the Guard's officer corps.

Table A.1
Female and Minority Representation in the Army National Guard and Reserve, FY 1990–2001

FY	Women				Racial/Ethnic Minorities			
	Army National Guard		Army Reserve		Army National Guard		Army Reserve	
	Officers	Enlisted	Officers	Enlisted	Officers	Enlisted	Officers	Enlisted
1990	7.4%	7.0%	20.3%	20.4%	10.9%	22.7%	16.0%	32.6%
1995	8.1%	8.2%	22.4%	22.4%	12.6%	27.0%	19.9%	42.3%
2000	9.4%	11.6%	24.8%	25.0%	14.0%	27.9%	25.3%	45.0%
2001	9.4%	12.4%	24.2%	25.0%	14.3%	28.2%	26.0%	45.3%
Percentage-point change (FY 2001–1990)	2.0	5.4	3.9	4.6	3.4	5.5	10.0	12.7

NOTE: Data include only members of the Selected Reserve.

SOURCE: Derived from Tables 4.14 and 4.23 in Office of the Deputy Under Secretary of Defense (Military Community and Family Policy), 2004.

for the air defense of the United States, as well as some close air support, airlift, and refueling aircraft, to the reserves. It also announced that the individual ready reserve would be used to offset active component shortages in transportation, comptroller, fuels, judge advocate, and weather functions. Additionally, the Air National Guard and Air Reserve began to take responsibility for a larger share of contingency and exercise missions (GAO, 1997).

Although the impact of force structure changes on demographic diversity is not clear, what is clear is that demographic diversity increased between FY 1990 and 2001 in the Air National Guard and Air Reserve. The last row of Table A.2 shows that female and minority representation increased anywhere from 2.8 percentage points (female enlisted in Air Force Reserve) to 5.1 percentage points (minority enlisted in Air Force Reserve) between FY 1990 and 2001.

Table A.2
Female and Minority Representation in the Air National Guard and Air Reserve, FY 1990–2001

	Women				Racial/Ethnic Minorities			
	Air National Guard		Air Force Reserve		Air National Guard		Air Force Reserve	
FY	Officers	Enlisted	Officers	Enlisted	Officers	Enlisted	Officers	Enlisted
1990	10.1%	13.6%	20.1%	18.8%	10.0%	16.8%	10.6%	25.7%
1995	12.6%	14.5%	24.1%	18.8%	10.5%	18.1%	11.3%	27.6%
2000	14.8%	16.7%	24.1%	20.5%	12.9%	21.2%	12.8%	30.0%
2001	14.9%	17.6%	24.6%	21.6%	13.0%	21.5%	13.3%	30.8%
Percentage-point change (FY 2001–1990)	4.8	4.0	4.5	2.8	3.0	4.7	2.7	5.1

NOTE: Data include only members of the Selected Reserve.
SOURCE: Derived from Tables 4.14 and 4.23 in Office of the Deputy Under Secretary of Defense (Military Community and Family Policy), 2004.

Naval Reserve and Marine Corps Reserve

The 1997 QDR recommended that the Navy and Marine Corps cut more than 4,000 reserve forces apiece. The Navy achieved about half of these cuts through reductions in force structure, in particular, by shrinking the size and number of its reserve P-3 maritime reconnaissance squadrons. It took care of the rest by eliminating positions in reserve support activities and funded positions, such as underwater construction, that it had been unable to fill. Following an internal force structure review, the Marine Corps Commandant at the time decided to eliminate 3,000 reserve positions. Less than half of these came from drilling units, whose personnel belonged to sites being deactivated or realigned. The rest were individual mobilization augmentees and reservists on active duty. By eliminating a substantial number of relatively high-cost full-time reservists, the Marines were able to achieve approximately the same level of savings and lose fewer personnel than directed in the QDR. The Marine Corps also cut back on the number of new recruits (GAO, 1998).

As with the Army and Air Force Reserves, the Navy Reserve and Marine Corps Reserve experienced increases in demographic diversity during the 1990s. Table A.3 provides similar information as Tables A.1 and A.2 but for the Navy Reserve and Marine Corps Reserve. As the last row shows, female and minority representation increased, especially in the Navy Reserve. The largest increase was for enlisted minorities in the Navy Reserve, who made up less than 20 percent of the force in 1990 but over 35 percent by 2001.

Summary

Overall, the post–Cold War reserve force reductions were significant—mainly, in the Army—but not as steep as they were in the active component. In part, this was due to the National Guard's ability to stave off larger cuts in Congress and, in part, reflected DoD's intent to shift

Table A.3
Female and Minority Representation in the Navy Reserve and Marine Corps Reserve, FY 1990–2001

| | Women | | | | Racial/Ethnic Minorities | | | |
| | Navy Reserve | | Marine Corps Reserve | | Navy Reserve | | Marine Corps Reserve | |
FY	Officers	Enlisted	Officers	Enlisted	Officers	Enlisted	Officers	Enlisted
1990	14.3%	15.3%	4.9%	3.9%	9.3%	16.7%	11.2%	26.5%
1995	16.2%	17.5%	5.9%	3.7%	13.3%	25.8%	8.8%	29.8%
2000	16.9%	20.0%	6.0%	4.6%	20.2%	34.3%	11.1%	33.0%
2001	17.0%	20.6%	5.8%	4.7%	22.0%	36.3%	11.7%	33.5%
Percentage-point change (FY 2001–1990)	2.7	5.3	0.9	0.8	12.7	19.6	0.5	7.0

NOTE: Data include only members of the Selected Reserve.
SOURCE: Derived from Tables 4.14 and 4.23 in Office of the Deputy Under Secretary of Defense (Military Community and Family Policy), 2004.

certain missions and elements of force structure from the active component to the reserve component during a period of reduced security threat. The Army's support units experienced the brunt of the reserve force drawdown. Although our information is limited with respect to the Air Force, Navy, and Marine Corps reserve components and the Army Reserve, female and minority representation generally increased across all services' RCs during the 1990s reductions.

Recent Drawdown

Between 2001 and 2013, the overall authorized end-strength of the selected reserves dropped from 874,664 to 850,880, an approximately 3-percent decline. During this period, the largest shifts in authorized end-strength occurred in the Army National Guard (up 2.2 percent), the Air Force Reserve (down 4.7 percent), and, particularly, the Navy Reserve (down 29.7 percent). A smaller change occurred in the Air National Guard (down 2.1 percent), while the authorized end-strengths of the Army Reserve and the Marine Corps Reserve remained largely unchanged. In the near term, the only reserve components expected to face significant reductions are the Navy Reserve and the Army National Guard, which had been programmed to lose 3,400 and 4,000 personnel, respectively, in 2014 (Jansen et al., 2014). Although the longer-term outlook for reserve forces is less clear, RC advocates contend that DoD should make it as easy as possible for those leaving the active component due to the drawdown to join the reserves (Daniel, 2012).

The remainder of this section details the approaches of the various RCs to recent drawdowns as well as the potential impact of the drawdown on the reserve forces.

Army National Guard and Reserve
Continuing Debate Over Active-Reserve Force Mix
The debate over the appropriate mix of active and reserve forces has continued during the recent drawdown, with arguments reminiscent of those made in the 1990s made on both sides of the issue. For his

part, Army Chief of Staff General Raymond T. Odierno stated that the Army would have to increase its dependence on the ARNG and USAR to compensate for serious losses in active forces, "particularly if the United States gets into two major long-term combat operations at the same time." However, he has questioned whether active and reserve forces were "interchangeable" and whether the latter were cheaper than the former when mobilized (Feickert, 2014a, p. 7). As during the post–Cold War period, reserve force skeptics continue to emphasize the difficulties of training and deploying reserve combat units in emergency situations (Rostker, 2013). Apparently sharing this perspective, Odierno has stated that the active component's proportion of the Army should not fall much below its level during the wars in Afghanistan and Iraq (51 percent), which may require additional cuts to the National Guard. For their part, National Guard advocates seem prepared to fight further cutbacks, believing they can persuade Congress to fund the Guard at its current end-strength (Jordan, 2014).

Current State of Play

Having to move forward with its planning process, the Army in early 2014 proposed to reduce the National Guard and Reserve by about 10 percent over six years ("DoD Makes It Official: Budget Cuts Will Shrink Army to 420,000 Soldiers," 2014). According to Army National Guard experts we interviewed, Guard officials developed a force structure divestiture plan that the Army finds acceptable. If necessary, the plan would reduce combat, aviation, and enabler forces from the Guard's overall end-strength of 358,000 in 2013 to 335,000 in 2017 and 315,000 in 2019. According to Army Reserve experts we interviewed, the Army Reserve currently faces a different personnel situation. Army Reserve officials expect to fall about 10,000 short of their end-strength objective of 205,000 in 2014. In the short term, they are desperately trying to grow the reserves so as not to lose programmed funding. In the longer term, they may need to shave off about 10,000 personnel by 2019 if they are to reach the 185,000 target designated by the Budget Control Act.

Positive Consequences of the Drawdown

Despite possible reductions, the Army's reserve components stand to benefit from the larger drawdown to the active component. To encourage these transitions from active to reserve, reserve recruiters are allowed to meet with departing active soldiers earlier than before (Lopez, 2014). Additionally, a pilot program launched at Fort Hood, Texas, in the winter of 2014 offers expanded incentives for joining the Guard or Reserve, to include:

- Bonuses for soldiers with high-demand skills
- The chance to leave active duty up to a year early
- More opportunities to become a warrant officer
- Chances to retrain for a new job while still serving out their active-duty contract.[3]

By stepping up their efforts to attract qualified and experienced transitioning soldiers, the reserve components are hoping to fill critical shortages in their ranks. In particular, the USAR has vacancies at the midgrade level within its noncommissioned officer corps, which it hopes to fill with former members of the active component.

The ARNG is also interested in recruiting troops coming off active duty. According to General Frank Grass, chief of the National Guard Bureau, about 50 percent of new ARNG soldiers had prior military service before the wars in Iraq and Afghanistan. In recent years, however, the number of new ARNG soldiers with prior military service has shrunk to about 20 percent, which has meant more money for training and less operational experience in the force. Gass hopes to use the drawdown to alter this trend (Tan, 2014).

Reserve Personnel Reduction Strategies

At the time this report was written, neither the ARNG nor the USAR announced the particulars of their current drawdown strategies. ARNG officials we interviewed indicated that they would use a transparent process and, in cooperation with the states' adjutants general,

---

[3]   Reportedly, if this incentive program proves successful, it would be implemented Army-wide (Tan, 2014).

would "determine the best force that can be implemented" either at the 335,000-force level specified by the 2014 Budget Control Agreement or the 315,000 level indicated by the 2011 Budget Control Act. In deciding which units to eliminate, ARNG will compare the performance and readiness of similar units over a five-year period so none will be advantaged or disadvantaged as a result of recent deployments. To meet the emergency needs of the states and maintain a sufficient national reserve force, no state will lose more than one brigade headquarters without a replacement.

During the process of shaping the future force, ARNG officials anticipate that significant re-balancing will be required for states facing force structure losses. Unlike active component soldiers periodically transferred to different locations within and outside the United States, Guard soldiers typically remain within their state borders. If after a drawdown no compatible unit remains within reasonable commuting distance, soldiers may elect to separate from the Guard, reclassify in order to complete their careers, or transfer to the USAR if they can find a suitable military occupational specialty.

As mentioned above, the USAR is currently less concerned with managing reductions than with growing end-strength. In the summer of 2014, Army reserve personnel fell 9,000 short of the number programmed for that year: that is, close to the 195,000 established for 2017. Despite the current overall shortfall, the USAR has an overage of junior enlisted soldiers. To deal with this problem and meet future drawdown requirements, one USAR official recommended either restricting accessions, transferring personnel into the Individual Ready Reserve, or eliminating from the service those who fail to appear for drills or prove unsatisfactory for some other reason, such as not achieving height, weight, or physical training standards.

Demographic Impact of Recent Drawdown

Based on our interviews with ARNG and USAR officials, neither the ARNG nor USAR has taken a close look at the demographic impact of recent force reductions. That said, ARNG personnel data show that the percentages of minorities and women within the ARNG rose, while the percentage of white males fell during the past several years.[4] At the start of FY 2010, blacks, Hispanics, and other racial-ethnic minorities constituted 13.6 percent, 8.6 percent, and 4.4 percent of the Guard's enlisted population, respectively, whereas whites composed 73.3 percent. By the end of July 2014, the percentages of black, Hispanic, and other minority enlisted Guard personnel had increased to 15.5 percent, 10 percent, and 5 percent, respectively; the enlisted white population declined to 69.6 percent. The same pattern held for ARNG officers, but to a lesser extent. The percentage of whites in the officer corps slipped by a little over 1 percent between 2010 and mid-2014 (from 82.3 to 81.1 percent), while the populations of black, Hispanic, and other minority officers each grew by less than 1 percent. The share of women in the ARNG increased slightly from 14.5 percent to 15.3 percent of enlisted soldiers, and from 12.4 percent to 12.7 percent of officers between 2010 and 2014.

Although the drawdown's impact up to mid-2014 on minority and female populations could be seen as negligible or even positive, this positive picture is somewhat diminished by the data on enlisted administrative separations. These separations include those due to reductions in authorized end-strength as well as separations due to unsatisfactory participation and failure to meet body composition standards, among other reasons. As the first cell in the last row of Table A.4 shows, almost 19 percent of the enlisted administrative separations from FY 2010 to the end of the third quarter of FY 2014 were meted out to black Guard personnel, although they constituted, on average, over 14 percent of the enlisted population within the ARNG during this period. Black Guard personnel were administratively separated at higher rates than other racial/ethnic groups in most occupational categories, but especially in the force sustainment job category. Black personnel composed more than a quarter of those separated in this category, which included almost 40 percent of all administrative separations.

As Table A.5 indicates, female enlisted members of the Guard were administratively separated relatively more than male enlisted members

---

[4]   These data were provided by Headquarters, Army National Guard, Personnel, Programs and Management Division in July 2014.

Table A.4
Percentages of Enlisted Administrative Separations (and End-Strength) by Occupational Category and Racial/Ethnic Group, FY 2010 Through Third Quarter of FY 2014

Occupational Category	Racial/Ethnic Group				Total
	Black	Hispanic	Other Minority	White	
Force Sustainment	10.1 (7.7)	3.4 (3.9)	1.5 (1.8)	24.7 (24.2)	39.7 (37.5)
Health Services	0.7 (0.5)	0.3 (0.4)	0.2 (0.3)	3.2 (3.7)	4.5 (4.9)
Operations	4.2 (3.2)	2.6 (3.2)	1.3 (1.6)	25.9 (27.5)	34.0 (35.5)
Operations Support	1.4 (1.1)	0.5 (0.7)	0.4 (0.4)	5.1 (5.6)	7.4 (4.8)
Other	0.9 (1.1)	0.6 (0.7)	0.4 (0.4)	4.6 (5.9)	6.4 (8.0)
Unassigned	1.4 (1.1)	0.6 (0.7)	0.3 (0.4)	5.8 (5.9)	8.0 (8.0)
Total	18.8 (14.3)	8.0 (9.3)	4.0 (4.7)	69.3 (71.7)	100.0 (100.0)

SOURCE: Headquarters, Army National Guard, Personnel, Programs and Management Division.

Table A.5
Percentages of Enlisted Administrative Separations (and End Strength) by Occupational Category and Gender, FY 2010 Through Third Quarter of FY 2014

Occupational Category	Gender Group		Total
	Women	Men	
Force Sustainment	10.3 (8.8)	29.4 (28.7)	39.7 (37.5)
Health Services	1.7 (1.4)	2.8 (3.5)	4.5 (4.9)
Operations	2.4 (2.1)	31.6 (33.4)	34.0 (35.5)
Operations Support	1.6 (1.3)	5.8 (6.5)	7.4 (4.8)
Other	1.1 (1.0)	5.4 (5.4)	6.4 (8.0)
Unassigned	1.0 (0.8)	7.1 (7.2)	8.0 (8.0)
Total	18.0 (15.3)	82.0 (84.7)	100.0 (100.0)

SOURCE: Headquarters, Army National Guard, Personnel, Programs and Management Division.

over the last few years. Almost 18 percent of enlisted soldiers receiving administrative separations from FY 2010 to the end of the third quarter of FY 2014 were women, although they constituted on average about 15.3 percent of the enlisted corps. Similar to occupational trends for black personnel, female separations were more prevalent in the force sustainment category than in other occupational categories.

Air National Guard and Air Reserve

As in the Army, most of the recent Air Force personnel cuts have occurred in the active component. In part, this is a consequence of drawdowns as active Air Force members join the Guard and Reserves in order to retain affiliation with the service. However, it also stems from the political influence of reserve force advocates in Congress. For example, in 2013, Congress rejected DoD's proposal to significantly reduce the size of the Air National Guard and Air Reserve in accordance with its plans to divest, transfer, or retire certain aircraft within the reserve component (DoD, 2013). In the end, lawmakers authorized only a small reduction in end-strength for the Air National Guard (from 106,700 to 105,700) and the Air Reserve (from 71,400 to 70,880) (Jansen et al., 2014).

When Air National Guard personnel are affected by force-shaping measures, the Air Force has several ways to "soften the landing." For example, it can offer early retirement to full-time active Guard and Reserve personnel (not traditional guardsmen), at least for the time being. In addition, the service can pay for Guard personnel willing to move to another state if their unit is facing reduction, and Guard wings not facing reductions can be asked to take personnel in wings that are facing reductions. Finally, the Air Force may allow some Guard units to maintain excess capacity for about two years to handle the influx of Guard personnel from downsized units.

As of fall 2013, the Air Reserve is less focused on force reductions than on retaining skilled personnel and filling gaps in certain mission areas. For the Air Reserve, unit relocations—such as those due to Base Realignments and Closures (BRACs)—have a major negative impact on retention. (The reserves tend to lose people when a unit is moved from one state to another.) Retirement-eligible reservists, in particular, are more likely to leave the service than move to a new location. During force-reduction periods, the Air Reserve has the authority to offer incentives to induce traditional reservists with critical skills and in certain locations to relocate to locations where their skills are needed. Incentives include inactive duty training pay, Permanent Change of Station authorities, and educational benefits.

Air Force officials did not provide detailed information on the demographic implications of limited reserve component reductions. However, they indicated that the recent drawdown might have a disproportionate impact on minority and female reservists because they tend to be located in support career fields targeted for cuts.

Naval Reserve

To help reduce the strain of the wars in Iraq and Afghanistan and better integrate its reservists into active-duty operations, the Navy conducted a zero-based review of the Naval Reserve's existing and required skills in 2004. As a result, the Navy increased its reliance on the reserves to compensate for the drawdown in its active forces while at the same time reducing certain parts of the Navy Reserve (Kennedy, 2004). In 2005, the Naval Air Forces announced plans to eliminate 3,120 reserve positions (GAO, 2005). In 2013, the Navy said it would cut reserve end-strength by approximately 2,500 over five years because of decreased demand for expeditionary combat assets, such as cargo handling and construction units. This involved the elimination of six reserve Naval Mobile Construction Battalions as well as 45 percent of the Seabee billets that had existed previously (U.S. Navy Reserve, 2012).

Like the Air Force officials we interviewed, Navy officials did not provide specific information on the demographic consequences of recent reserve reductions. However, Navy officials did not believe that recent cuts in aviation squadrons, cargo handling battalions, and construction units would involve disproportionate numbers of women or racial/ethnic minorities.

Marine Corps Reserve

According to Marine Corps officials we interviewed, in 2010, the Marine Corps conducted a thorough force structure review that identified capabilities that ought to be retained in the active and reserve components. To improve operational efficiency and reduce personnel costs, the review proposed placing the reserve division, wing, and logistics group headquarters in cadre status and eliminating the Mobilization Command headquarters (U.S. Marine Corps, 2011). Despite this, the overall size of the Marine Corps Reserve has remained about the same for more than a decade and is expected to come down by only 1,100 in the near term (from 39,600 to 38,500). Moreover, the reserves are projected to grow in certain areas as they downsize in other areas. To address midcareer vacancies in its reserve force, the Marine Corps established a number of programs designed to transition midcareer enlisted personnel (E-5s and E-6s) from the active force to the reserve force.

It is possible that the number of Marine reservists could actually increase in the future. According to one defense force structure expert, if Marine Corps leaders decided to transform the service into "a more expeditionary, crisis response type of force, many of the capabilities needed for fighting a major land war could be shifted to the reserve component." Such capabilities could include tank, artillery, and fixed-wing aviation units (Feickert, 2014b).

As of April 2014, the Marine Corps Reserve's drawdown strategy is to reduce enlisted accessions. As just indicated, there is an abundance of junior Marines in the reserves covering for gaps in the middle ranks. As those leaving the active component fill these gaps, the Marine Corps intends to hold down the number of E-1s through E-3s in the reserve force. However, the Marine Corps Reserve's officer accessions strategy will likely remain unchanged.

Although uncertain of the demographic effects of its proposed drawdown, Marine Corps Reserve officials note that about 20 percent of its enlisted population is Hispanic. Consequently, if initial enlisted accessions are reduced, there will probably be a corresponding dip in the number of Hispanic marines in the reserve force. That said, the small size of the reduction would not likely have a large impact on Hispanic representation.

Summary

The Navy Reserve has been the only reserve component to experience a significant drawdown in recent years. Although other reserve components—in particular, the Army National Guard—are downsizing, the extent of future reductions is somewhat unclear due to the uncertainty over sequestration, the political influence of the Guard, and the outcome of the active-reserve force mix debate within DoD. It is likely, however, that reserve force reductions will not only be smaller than those for the active force, but also significantly less than the reserve cutbacks of the 1990s. With the exception of the Navy Reserve, all of the reserve components are aggressively courting midgrade personnel leaving active service, and RC officials indicate that they can deal with any needed reductions through a combination of natural attrition, reduced accessions, transfers to the Individual Ready Reserve, and administrative separations. The data on the impact of recent personnel reductions on the reserve workforce are limited. While the size of the minority and female populations within the ARNG has grown in recent years, black and female Guard personnel have been disproportionately affected by administrative separations.

Comparing the Two Drawdown Eras

How do the consequences of personnel drawdowns for DoD's reserve forces compare across time, service, and component? Table A.6 shows a number of gaps in our knowledge regarding the specific strategies that the services employed to shape their reserve components in the post–Cold War period as well as the potential effects of the recent drawdown on the demographic composition of the reserve forces. However, available evidence indicates that the consequences of neither drawdown were as severe for the reserves as they were for the active components. That said, the cutbacks in the 1990s were significantly larger than those that have taken place in recent years, although, in both time periods, reductions have hit some service reserve components harder than others; the Army RCs were hit during the post–Cold War drawdown, and the Navy Reserve in recent years. The services' strategies for reducing

Table A.6
Comparison of Reserve Component Reductions in the 1990s and 2010s

Reserve Component	Extent of Reductions		Reduction Strategy		Diversity Changes	
	1990s	2010s	1990s	2010s	1990s	2010s
ARNG	Substantial	Small	?	Force structure changes, voluntary separations, transfers	Small-to-moderate increases	Small but with some decreases for enlisted
USAR	Substantial	Small	?	Restricted accessions, transfers, involuntary separations	Moderate-to-large increases	Possibly small
Air Guard	Small	Very small	?	Force structure changes, early retirements	Small-to-moderate increases	Possibly very small
Air Force Reserve	Small	Very small	?	BRACs, voluntary separations, transfers	Small-to-moderate increases	Possibly very small
Naval Reserve	Small	Substantial	Force structure changes, positions eliminated	Restricted accessions	Moderate-to-large increases	No significant changes expected
Marine Reserve	Small	Very small	Eliminated positions, restricted accessions	Force structure consolidations, restricted accessions	Small increases	Possibly small

reserve forces have varied somewhat across components and drawdown eras. Restricting accessions appears to be the most common method of meeting the reduction goals of the current drawdown period. Force structure changes constitute key elements of the force-shaping strategies of the two National Guard components (and the Naval Reserve), but not the Army and Air reserves.

Because of data limitations, it is difficult to compare the impact of the two drawdowns on the composition of reserve workforces. The personnel data suggest that female and minority representation generally increased across the reserve components in the 1990s, with some exceptions. Although we have limited data on the demographic impact of the recent reductions, ARNG data show that enlisted black and female personnel have been disproportionately subject to administrative separations, possibly, in part, because they tend to be concentrated in support occupations targeted for downsizing. It is not clear what this trend in the ARNG augurs for the future as the drawdown gathers steam. However, the extent to which reserve force reductions are focused on support units and new accessions (which tend to be more diverse than the overall military population) could determine the impact that downsizing efforts have on demographic diversity within the reserves.

Civilian Drawdowns

The purpose of this appendix is to compare the defense-civilian personnel drawdowns within the Army, Air Force, and Navy from the end of the Cold War to the present in order to better understand factors affecting demographic diversity of the DoD civilian workforce[1] during drawdowns. Relying on interviews with civilian personnel officials in the three military departments, limited aggregated personnel data on direct hire civilians, and open source and DoD documentation on civilian personnel policies and effects, this appendix provides a largely qualitative and descriptive account of DoD's attempt to shape its civilian workforce over the last 25 years.

Like Appendix A, this appendix has three major sections. The first focuses on the 1990s drawdown, and the second focuses on the recent drawdown. Each section addresses the following four personnel management topics:

1. the principal drivers of change in the size and shape of DoD's civilian workforce
2. the nature, scope, and timing of civilian personnel changes within each military department
3. measures taken by DoD to effect desired personnel changes within its civilian workforce
4. actual changes in the composition of civilian workforces during the drawdown period.

---

[1] Our review does not focus on DoD agencies (e.g., Defense Logistics Agency) outside the services.

The appendix concludes with a comparison of DoD responses to civilian personnel drawdowns—and their impact—across time and service.

Post–Cold War Era

This section addresses the principal drivers of the post–Cold War DoD civilian reductions, the size and scope of the reductions, reduction strategies and approaches, and the consequences of reductions, particularly with respect to demographic diversity.

Principal Reduction Drivers

Several important studies and legislative actions underlay the downsizing that took place in the civilian workforce in the decade after the end of the Cold War. In its 1993 Bottom-Up Review of national defense strategy and resource requirements, DoD found that civilian personnel not only constituted a significant part of infrastructure costs, but civilian workforce cutbacks trailed behind military personnel reductions. As a result, the department decided to downsize the civilian workforce in line with the military and to minimize civilian-related infrastructure costs over a period of six years. This downsizing trend was reinforced by broader initiatives affecting all federal government civilians. In September 1993, for example, the White House's National Performance Review recommended that federal civilian workforce be restructured by

- concentrating civilian downsizing among supervisors, headquarters staff, personnel specialists, budget analysts, procurement specialists, accountants, and auditors
- doubling federal agencies' current ratio of one manager or supervisor for every seven employees to a ratio of one to 14 by 1999
- reengineering, or reinventing, government through streamlining to achieve personnel and fiscal savings.

In response to this review, DoD announced a plan to cut its civilian employees by 18 percent, exceeding the White House's recommended 12-percent reduction (GAO, 1996). In 1994, Congress passed the Federal Workforce Restructuring Act, which placed annual ceilings on executive branch full-time equivalent (FTE) positions for fiscal years 1994 through 1999 and resulted in the elimination of about 200,000 federal jobs (Bowling, 1996).

Size and Scope of the Drawdown

In the end, the post–Cold War reduction in the DoD civilian workforce was considerably larger than initially envisioned. Between fiscal years 1989 and 1999, the number of full-time positions declined by about 400,000, from approximately 1,117,000 to 714,000—a 36-percent reduction (Brostek and Holman, 2000). This represented the bulk of federal civilian jobs lost during the 1990s (Schwellenbach, 2013). Cuts were distributed relatively equally throughout the services. By the end of the 1990s, for example, about 96,000 fewer civilians worked for the Air Force than at the beginning of the decade (approximately 153,000 versus 249,000), a 39-percent drop (Chapman, 1996). In contrast to the military departments, civilian employment in the defense agencies actually increased in the post–Cold War period. Because of consolidations and transfer of some functions from the services, the number of full-time civilians in the defense agencies grew by 48,000 between 1987 and 1997 (Holliman, 1993).

Reduction Strategy and Approach

According to the GAO (1996), DoD civilian reductions during the 1990s were not guided by a comprehensive management or downsizing strategy. Instead, OSD and service headquarters relied on commanders and managers of defense organizations to determine their minimum skill and staffing requirements. Their primary role was to establish civilian workforce reduction targets and monitor progress in meeting those targets. While subordinate commanders were told to reduce the civilian workforce through attrition if possible, they were given certain tools to maximize internal placement opportunities, retain critical skills, and shape the composition of the civilian workforce. Tools included monetary incentives to encourage voluntary separation of employees in surplus skill categories and areas undergoing major reductions in force (DoD, 1993). Despite these measures, GAO concluded, "DOD's approach to civilian force reductions was less oriented toward shaping the makeup of the workforce than was the approach it used to manage its military downsizing" (Brostek and Holman, 2000, p. 7).

In the early years of the post–Cold War drawdown, DoD's principal downsizing methods were voluntary attrition and retirements and freezes on hiring authority (Brostek and Holman, 2000). For example, the Army executed a civilian hiring freeze that was even more stringent than the DoD-wide freeze in order to meet its target of 123,000 reductions between 1987 and 1997. According to this policy, a single hire was permitted from the outside for every four losses to the department (Holliman, 1993). DoD officials recognized that hiring freezes and generalized attrition were problematic reduction tools because they made it difficult to downsize in an orderly manner, achieve reductions when and where they were needed, and fill essential positions when vacancies occurred (GAO, 1993). Therefore, targeted attrition and reductions in force were used to a limited extent as deliberate shaping measures—in addition to BRACs and transfers to defense agencies.

Voluntary Separation Incentives

The 1993 NDAA authorized a number of transition assistance programs for civilian employees, including financial incentives to civilians who voluntarily retired or resigned. According to this legislation, those eligible could receive whichever was less: either $25,000 or the amount of severance pay to which they were entitled (Holliman, 1993). In some cases, separation incentives were targeted to specific positions, while in other cases they were offered to large groups of employees in order to meet installation reduction goals. In a 1993 assessment of the use of these incentives, DoD civilians had requested and been approved for between 70 and 80 percent of the incentives authorized, the majority separating under provisions for regular or early retirement (GAO, 1993).

Reductions in Force

Although meant as a tool of last resort, the services did make use of reductions in force when voluntary measures failed to achieve the desired level of civilian reductions. To cushion the impact of these separations, the department required 120 days' notice of intent to lay off an employee if there were more than 50 such actions at one base. This rule, unique to DoD, provided opportunities to take advantage of early retirement or separation incentives or to hunt for jobs in other parts of the department. For example, in 1994, when Air Force officials at Hill Air Force Base, Utah, determined they might have to lay off 800 civilian employees, they conducted a "mock RIF" to let people know who might be in line to lose their jobs. By engaging early with employees and using alternative separation methods, Hill Air Force Base managed to cut its potential RIFs from 800 to 264 (Chapman, 1996). Across DoD, relatively few civilians lost their jobs through RIFs in the 1990s. According to one defense news article, RIFs accounted for fewer than 9 percent of the 288,000 DoD civilian positions eliminated from 1989 through 1995 (Gillert, 1996).

Whether the limited use of RIFs during the 1990s was good or bad is debatable. Targeted layoffs can be a useful means of maintaining the appropriate balance of skills during a drawdown. However, civilian workforce regulations make it difficult for DoD to safeguard employees with the right set of skills and jettison those without them. Title 5 of the U.S. Code stipulates that federal departments must take into account veteran status, years of service, and performance of employees when making layoff decisions. According to at least one civilian personnel expert we interviewed, nonretired veterans with seniority are typically the last employees to be let go, whereas nonveteran junior employees are the most vulnerable to RIFs. Consequently, a RIF can exacerbate rather than alleviate skill imbalances. This occurs when senior employees, having prior experience in lower-graded positions, "bump" junior employees out of their positions even though they may not meet the current skill requirements of the positions into which they are "retreating" (GAO, 1993).

Priority Placement Program

During the post–Cold War drawdown, the primary tool to assist DoD civilian employees in danger of being laid off was the Priority Placement Program (PPP), which is an automated, worldwide referral service that matches employee skills with vacant positions elsewhere in the department. Unlike other placement assistance programs, PPP guarantees that registrants whose job qualifications match job requirements of vacant positions must be given an offer of employment. PPP's effectiveness in placing excess employees varied during the early years of the drawdown. From 1989 through 1992, between 24 percent and 66 percent of high-priority candidates received job offers or were placed in other positions (Holliman, 1993).

BRACs and Transfers

Although not strictly personnel reduction methods, BRAC actions and transfers to defense agencies were used extensively by the services in the 1990s to help achieve their downsizing targets. For example, in 1996, the Navy estimated that BRAC actions would yield about 35 percent of the 90,500 Navy and Marine Corps civilian reductions targeted for 1993 through 2001. For its part, the Army expected 14 percent and the Air Force expected 9 percent of planned civilian reductions to result from BRAC actions (GAO, 1996). In addition, much of the 1989–1995 reduction in the Air Force's civilian workforce was reportedly accomplished through the shift of personnel from service to DoD agencies, such as the Defense Finance and Accounting Service and the Defense Commissary Agency (Chapman, 1996).

Impact of Reductions on the Civilian Workforce

As might be expected, the 1990s drawdown affected civilian workers in some job categories more than in others. The largest reductions were in clerical positions and blue-collar jobs (i.e., wage-grade employees) (DoD, 2000). The Army's 1993 annual review of its civilian workforce stated that blue-collar employment had dropped 39 percent and clerical employment dropped 26 percent over the previous year, while professional and administrative employment had dropped only 6 percent (GAO, 1993). According to DoD's Acquisition 2005 task force report, the civilian acquisition workforce was about half the size it had been ten years before, and it predicted that half of the employees in the acquisition force at the time would have retired by 2006 (Walker, 2003).

DoD's reliance on hiring constraints and voluntary attrition to achieve civilian reductions in the 1990s led to demographic, as well as skill, imbalances. In particular, the drawdown resulted in an older civilian workforce. The median age of DoD civilian employees rose from 41 in 1989 to 46 in 1999, and the number of civilians under the age of 31 dropped by 76 percent during the same period, while the number of those aged 51–60 remained about the same. This aging process, which occurred across the occupational spectrum, inevitably resulted in a more experienced workforce. Between 1989 and 1999, the median length of service increased from 11 to 17 years, while the number of civilians with less than five years' experience dropped by 69 percent (DoD, 2000). In the view of GAO, the "graying" of the civilian workforce threatened DoD's ability to generate "new and creative ideas and develop the skilled civilian workers, managers, and leaders it will need to meet future mission requirements" (Walker, 2003, p. 4).

The racial/ethnic and gender consequences of the drawdown on DoD's civilian employees have not been definitively determined. However, some indication of demographic impact can be found in GAO's review of downsizing results at three installations (Navy, Air Force, and Army) between 1991 and 1993. According to this report, minorities were involuntarily separated at a rate disproportionate to their numbers in the overall civilian workforce at all three locations, and women were separated in disproportionate numbers at two locations. In some cases, these results were due to minorities and women lagging behind nonminorities and men with respect to tenure, veteran's preference, or seniority. In other cases, the disproportionate separations occurred because many of the positions abolished at the installations belonged to minority employees who had no assignment rights to other positions. To lessen the potential negative effects of the drawdown on minority and female representation in DoD's civilian workforce, GAO recommended that DoD make greater use of separation methods that encourage retirements, thus enticing more white men, who tended to be older and have more job tenure than women and minorities, into voluntarily leaving the department (Kingsbury, 1994).

Summary

In sum, DoD experienced an almost 40-percent decline in its civilian workforce in the decade following the end of the Cold War. Defense civilian reductions were primarily driven by the political imperative of lower military spending during an era of relative peace as well as by Congress's desire that civilian employment come down at a rate equivalent to the overall military drawdown. Cuts were distributed approximately equally across the services. Although not guided by a comprehensive strategy, the service commands that implemented civilian reductions followed certain basic rules: use hiring freezes and voluntary separation measures first, reductions in force as a last resort. The impact of the 1990s civilian drawdown was felt most by particular occupational and demographic groups: clerical and blue-collar workers, junior employees, and, less certainly, black and female personnel.

Recent Drawdown

This section examines the motives, numbers, strategies, and effects associated with DoD civilian workforce reductions in recent years.

Principal Reduction Drivers

Recent DoD civilian workforce drawdown decisions are driven by the discretionary spending cuts required by the Budget Control Act (BCA) of 2011 as partially (and perhaps temporarily) ameliorated by the 2014 Bipartisan Budget Act (BBA). The first round of sequestration in March 2013 resulted in furloughs from a few days to almost two weeks for hundreds of thousands of DoD civilians (Reilly, 2013). Although DoD officials hope to avoid the organizational and personal disruption caused by furloughs in the future, permanent cuts in the civilian workforce are expected; the only question is how severe they will be. In order to address Congress's demand for significantly lower defense spending and needed modernization investments, the FY 2014 NDAA directs DoD to reduce headquarters management, including civilians and contractor personnel (Public Law 113-66, 2013). The only exceptions to these requirements appear to be certain high-profile initiatives such as support for Navy carriers, sexual assault prevention and veterans work programs, cybersecurity missions, and the Special Operations Command (DoD, 2013). As noted by a civilian personnel official we interviewed, proposed legislative guidance is for the current drawdown of DoD civilian and contractor workforces to be proportionate to the uniformed military drawdown.

Size and Scope of Reductions

Given lower discretionary budget levels and the continued threat of sequestration, the Army Chief of Staff, GEN Odierno, stated that the Army was preparing to reduce the end-strength of its civilian workforce by an estimated 14 percent (Odierno, 2013). Army officials we interviewed stated that the Army would be drawing down from a wartime high of 285,000 civilians in 2010 to 246,000 by 2015. Because of the 2013 hiring freeze, the Army was well under its programmed level of direct-hire civilians at the end of the first quarter of 2014 and was projected to remain so for the rest of the year.[2] Still, the Army continues to urge its commands to reduce civilian and contract labor in order to reach considerably lower 2016 civilian pay and end-strength targets.

Air Force civilian personnel officials appear less concerned about future cuts than their Army counterparts, asserting that Air Force civilian reductions are not expected to be significant. For the most part, Air Force civilian reductions would result from the divestment of weapons systems, such as the A-10 ground attack aircraft, or from weapons modernization initiatives, such as the transition from the C130J to the C130H aircraft, which requires less maintenance (and fewer civilian maintenance personnel). Likewise, Navy civilian personnel officials do not anticipate a substantial decline in its civilian workforce over the next several years. This is in part because the number of Navy civilians

---

[2] The bulk of DoD's civilian workforce is in the direct-hire civilian category. This category does not include foreign nationals employed by the department, civilians within the National Intelligence Program, or civilian labor provided by Overseas Contingency Operations funds.

grew by only about 10 percent between 2001 and 2013.[3] Thus, the Navy's 2012 total force analysis recommended reducing its approximately 200,000 direct-hire civilians by less than 2.5 percent or about 4,500 personnel.

Reduction Strategy and Approach

DoD intends to take a more strategic approach to downsizing the civilian workforce than it did during the post–Cold War era. According to former Secretary of Defense Chuck Hagel, about half of the civilian reductions in DoD depend on a new round of BRACs, the restructuring of military treatment facilities, and decreased demand for depot maintenance as the U.S. military transitions out of Afghanistan. Remaining reductions would result from workforce attrition and, if necessary, layoffs (Gore, 2013). As mentioned above, the Navy conducted a detailed review of its total force structure in 2012 to understand which functions grew in size and workload, the extent to which capabilities aligned with organizational structures, and areas where the service could take risks in terms of reductions. More recently, the Army launched a total force management initiative to understand the appropriate balance of military, civilians, and contractors within the service's generating force and ways to safely eliminate or decrease civilian costs associated with certain functions.

In addition to the civilian hiring restrictions in 2013, DoD leadership counseled the services to make "liberal use" of Voluntary Separation Incentive Pay (VSIP) and Voluntary Early Retirement Authority (VERA) in achieving civilian workforce reductions (Under Secretary of Defense for Personnel and Readiness, 2013). Whereas the latter is subject to certain eligibility requirements, VSIP of up to $25,000 can be given to federal employees not close to retirement age. Furthermore, VERA and VSIP can be taken separately or in combination, the latter being the most popular option. Those who opt for VSIP alone cannot return to federal service for at least five years on penalty of repaying

---

[3]   However, Navy officials noted that the dissolution of the Joint Forces Command in 2011 disproportionately affected the Navy civilian workforce compared to other services' civilian workforce.

VSIP (DoD Instruction [DoDI] 1400.25, 2009). Moreover, VSIP and VERA can be offered to offset involuntary separations. One purported advantage of voluntary incentives is that they can be used to reshape the workforce; they can target personnel in unnecessary managerial or supervisory positions or with low-demand skill sets to make way for personnel with high-demand skills (Under Secretary of Defense for Personnel and Readiness, 2013).

In part, DoD's current emphasis on voluntary incentives stems from the political imperative of avoiding widespread involuntary separations, but this does not mean that the administration is completely averse to laying off defense civilians. Secretary Hagel has stated his preference for targeted mandatory RIFs over a return to across-the-board civilian personnel furloughs (Schneier, 2013). Likely RIF targets include employees of OSD, military service headquarters, and combatant commands, all of which the 2013 strategic review recommended be cut by 20 percent (Reilly, 2013). If and when future civilian RIFs are announced, they will proceed in accordance with the same basic rules used during the post–Cold War drawdown. That is, human resources managers must create a "competitive area" to serve as the pool of jobs that could be eliminated. Then, administrators must take into account the employee characteristics described in the previous section in deciding whether an employee in the pool is retained or let go. Finally, employees whose jobs are abolished can—if qualified—bump colleagues in lower-graded positions. The bottom line is that the last to be hired tends to be the first to be fired. As in the 1990s, the department hopes to alleviate the plight of those facing RIFs through a range of civilian transition assistance programs, such as the previously described PPP (Reilly, 2013).

Army Reduction Methods

Following DoD's guidance, the Army is using VERA and VSIP as its principal tools for shaping the civilian workforce and RIFs only if subordinate commands cannot achieve their targeted cuts through other means. Army civilian personnel officials indicate that the service as a whole will receive about 8,000 VSIP authorizations for 2015. After receiving the commands' future personnel requirements and conducting predictive modeling, Army personnel officials will determine whether additional involuntary reductions in force may be needed. Although the Army has conducted RIFs in recent years, the number of involuntary separations from 2011 to mid-2014 has remained relatively low, reaching a high of over 1,000 in the final fiscal quarter of 2011 compared to about 10,000 voluntary separations and almost 4,000 retirements during the same period (see Figure B.1).

Originally, the Army planned for 2,155 civilian reductions in force in 2014. However, that projection was lowered to 989 RIFs because enough of the personnel slated for involuntary separation transitioned to other employment and less constrained BBA funding levels. As one Army civilian personnel official explained, civilian personnel hear that their organizations might be considering a RIF, so they think about leaving voluntarily before that happens. As a result, the number of RIFs turns out to be significantly lower than the number initially planned.

Figure B.1
Army Civilian Separations by Fiscal Quarter, 2011–2014

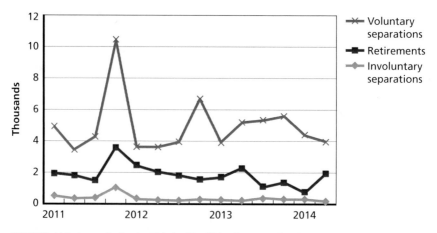

SOURCE: U.S. Army Assistant Chief of Staff for Personnel, AG-1-CP PAED.
NOTE: Includes only U.S. direct-hire employees funded by appropriations.
RAND RR1008-B.1

Air Force Reduction Methods

Like the Army, the Air Force has tried to use voluntary separation incentives as much as possible to meet the reduction demands of the current civilian drawdown. According to Air Force civilian personnel officials, in 2012, the Air Force had three rounds of VERA and VSIP for 3,476 personnel, two rounds in 2013 for 553 personnel, and one round in the first half of 2014 for 215 personnel. Although the Air Force has requested RIF authority in each of the last three years, the service involuntarily separated only 193 civilians in 2012, 64 in 2013, and none in the first half of 2014.

Navy Reduction Methods

According to Navy civilian personnel officials, the Navy has managed the current reduction somewhat differently from the Army and the Air Force. Rather than seeking broad-based solutions to downsizing, the Navy reviews each of its civilian positions in terms of its criticality and determines whether to continue to fill it. Although the Navy has enacted civilian hiring freezes in the past, the freezes are more akin to a "frost" that provides subordinate commands the flexibility to make additional hires when necessary. One Navy civilian personnel official acknowledged that the Navy uses authorized tools creatively to retain key personnel in the event of a base or facility closure; tools used include temporary hiring authorities and encouraging employees subject to a RIF to register early with the PPP. As a result, the number of involuntary reductions in the Navy is very small.

Impact of Reductions on the Civilian Workforce

Although data are not available to compare the impact of civilian reductions across the services, the limited evidence suggests that recent downsizing actions have had a somewhat disproportionate impact on certain occupational and demographic groups.

Army

According to Army personnel officials, members of the medical, acquisition, and maintenance communities have left the service in relatively large numbers in recent years. These officials claim that the predominant reason for these departures was that the furlough experience made continued Army employment seem unattractive to individuals with private-sector opportunities. However, certain occupations may be subjected to significant layoffs in the future. A recent RAND analysis found that the Army's Acquisition Support Command and the Network Enterprise Technology Command may require active cuts to civilian personnel unless internal transfer rates are lower than they have been in recent years (Nataraj et al., 2014). As Tables B.1 and B.2 show, from the start of 2011 until the end of the second quarter of 2014, the largest proportion of involuntary separations in the Army's civilian workforce was in the administrative (31.2 percent) and professional categories (25.2 percent).

The limited data provided by Army officials suggest that the impact of the Army civilian drawdown on women and minorities has been mixed. On the one hand, there has been little change in recent years in the proportions of racial/ethnic groups within the Army civil-

Table B.1
Percentages of Army Civilian Involuntary Separations (and End-Strength) by Occupational Category and Racial/Ethnic Group, FY 2011 Through Second Quarter of FY 2014

Occupational Category	Racial/Ethnic Group				Total
	Black	Hispanic	Other Minority	White	
Administrative	6.2 (6.2)	1.0 (2.0)	0.9 (1.5)	14.9 (21.4)	23.0 (31.2)
Blue Collar	3.5 (1.8)	1.0 (1.2)	0.5 (0.5)	14.1 (9.8)	19.2 (13.3)
Clerical	4.1 (2.6)	0.9 (0.8)	0.8 (0.6)	7.4 (5.5)	13.2 (9.5)
Other White Collar	3.7 (1.4)	1.0 (0.6)	1.0 (0.5)	9.9 (5.9)	15.7 (8.4)
Professional	1.8 (2.4)	0.4 (1.2)	1.0 (2.3)	9.2 (19.2)	12.4 (25.2)
Technical	5.5 (2.8)	0.9 (1.0)	0.8 (0.8)	9.1 (7.7)	16.5 (12.3)
Total	24.9 (17.2)	5.2 (6.8)	5.0 (6.2)	64.7 (69.6)	100.0 (100.0)

NOTES: Includes only U.S. direct-hire employees funded by appropriations.
Total includes personnel with unknown race/ethnicity (less than 0.3 percent of end-strength).
SOURCE: U.S. Army Assistant Chief of Staff for Personnel, AG-1-CP PAED.

Table B.2
Percentages of Army Civilian Involuntary Separations (and End-Strength), by Occupational Category and Gender, FY 2011 Through Second Quarter of FY 2014

Occupational Category	Gender Group		Total
	Women	Men	
Administrative	11.1 (12.3)	11.9 (19.0)	23.0 (31.2)
Blue Collar	1.8 (1.1)	17.4 (12.2)	19.2 (13.3)
Clerical	8.1 (6.6)	5.1 (2.9)	13.2 (9.5)
Other White Collar	3.2 (1.6)	12.5 (6.8)	15.7 (8.4)
Professional	5.7 (9.6)	6.7 (15.5)	12.4 (25.2)
Technical	10.7 (6.6)	5.8 (5.7)	16.5 (12.3)
Total	40.5 (37.8)	59.5 (62.2)	100.0 (100.0)

NOTES: Includes only U.S. direct-hire employees funded by appropriations.
SOURCE: U.S. Army Assistant Chief of Staff for Personnel, AG-1-CP PAED.

ian workforce. The percentage of black appropriated employees dropped slightly—from 17.5 percent to 17.2 percent of the total workforce—between 2011 and 2014.[4] During the same period, the Hispanic and white populations rose slightly (from 6.7 percent to 6.8 percent and from 69.5 percent to 69.6 percent, respectively) and the female population shrank by a small amount (38.2 percent to 37.8 percent). On the other hand, while relatively small in number, RIFs have hit black and female civilians the most. On average, black civilians were 17.2 percent of the workforce over the aforementioned period, but they represented 24.9 percent of those involuntarily separated.[5] By compari-

---

[4] The source for data in Tables B.1 and B.2 provided the information cited in this paragraph.

[5] We also looked at the results broken out by gender. Specifically, we conducted a chi-square analysis to compare the proportions of involuntary separations among black women, black men, white women, white men, women from other racial/ethnic groups, and men from other racial/ethnic groups. The results were statistically significant and indicated that black women and black men were disproportionately separated via administrative actions (e.g., 636 black women were separated, but only 444 would have been expected to be separated by chance). White women were not disproportionately separated. White men were underrep-

son, Hispanic civilians received 5.2 percent of involuntary separations, and white civilians received 64.7 percent, although they represented 6.8 percent and 69.6 percent, respectively, of total Army civilian end-strength. Black civilians were disproportionately subject to involuntary separation in four of the six major occupational categories: blue collar, clerical, other white collar, and technical (see Table B.1). Though 37.8 percent of Army civilian end-strength, female civilians received 40.5 percent of involuntary separations and were disproportionately laid off in the same occupational categories as black civilians (see Table B.2). Although the causes for these effects have not been established, it could be surmised that they are at least partly due to the fact that women and minorities tend to be less senior and thus more susceptible to RIFs. However, that does not explain why Hispanics have not been involuntarily separated at the same rate, or why black and female civilians have not experienced a disproportionate number of RIFs.

Navy and Air Force

Although Navy civilian personnel officials doubt that the recent civilian reductions have affected any particular demographic group more than others, Air Force civilian personnel officials have noticed a higher-than-normal separation rate among Hispanics, women, and persons with disabilities in recent years. Most Air Force separations have been voluntary and may not directly relate to the drawdown. However, Air Force officials speculate that minorities and women account for a disproportionate share of involuntary separations. This is because junior employees, who tend to be the most diverse component of the workforce, are more liable than older workers to be laid off, due to seniority rules. As Table B.3 shows, the clerical category, which is heavily female, experienced the largest number of involuntary reductions among Air Force civilian occupational categories from FY 2011 to 2012, followed by the administrative, technical, and blue-collar categories. Air Force civilian personnel officials also expressed concern about the continued aging of the civilian workforce. This process could be accelerated by

---

resented among those separated, and men and women from other racial/ethnic groups were slightly underrepresented among those separated as well.

Table B.3
Air Force Civilian Reductions in Force, by Occupational Category,
FY 2011–2012

Occupational Category	FY 2011	FY 2012
Administrative	44	10
Blue Collar	22	14
Clerical	79	8
Other White Collar	7	3
Professional	7	14
Technical	34	15
Total	**193**	**64**

SOURCE: Headquarters, U.S. Air Force, Deputy Chief of Staff, Personnel (AF/A1), July 2014.

the drawdown and, in particular, by the recent spate of civilian furloughs, which could cause junior employees to question the viability of a civil service career.

Veterans Preference in Civilian Hiring

The adverse impact of the veterans preference on federal government hiring of women and certain minorities is likely to be exacerbated by the current defense drawdown. According to a Syracuse University study (Lewis, 2013), veterans preference has substantially increased the percentage of federal employees who are men and probably decreased the percentages who are Asians, gay men, and immigrants. Given the demonstrated effects of past military reductions on federal employment, it is reasonable to expect that relatively few nonveterans will be hired during the next several years, thus altering the demographic balance of DoD's civilian workforce, especially with respect to women.

Summary

Although not likely to replicate the drawdown experience of the 1990s, the discretionary spending cuts associated with the 2011 BCA will probably result in substantial and permanent reductions in civilian employment within DoD. Although each of the military departments will face civilian cutbacks, the Army is expected to receive a larger share than the Air Force and the Navy. As in the post–Cold War era, the services have principally used voluntary separation incentives, early retirements, and selective hiring freezes to achieve their civilian reduction targets in recent years. While all have requested the authority to impose involuntary RIFs, the number of actual RIFs has remained low, especially in the Air Force and the Navy. Furthermore, the services have pledged to avoid a repeat of the 2013 civilian furloughs, which they worry may cause an exodus of highly skilled younger employees. The potential impact of the drawdown on the composition of the civilian workforce has not been examined in depth. However, Army and Air Force personnel data and interviews suggest that younger employees, minorities, women, and persons with disabilities in a range of occupations are disproportionately leaving (voluntarily and involuntarily) the DoD civilian workforce. Furthermore, the federal government preference for hiring veterans—in combination with the current military drawdown—could lead to a decreasing percentage of women and certain minorities in the ranks of DoD's civilian workforce.

Comparing the Two Drawdown Eras

As with the military side of DoD, across-the-board budget cuts were primarily responsible for the drawdowns of civilian personnel during the 1990s and in recent years. However, as one civilian personnel official noted, politics play a larger role in civilian workforce reductions than military force reductions. In part, this stems from the fact that, until recently, civilians have not been factored into DoD's strategic analysis of its total force requirements. In part, it results from congressional pressure to ensure that civilian and military drawdowns are commensurate. Finally, the size of DoD's permanent civilian workforce has been influenced by differing political views on the need to outsource various functions to private contractors.

Table B.4 summarizes the consequences of the two civilian drawdowns in terms of the size of reductions, the strategies used, and the impact of reductions on various components of the workforce. As the

Table B.4
Summary of Direct-Hire DoD Civilian Personnel Reductions During the 1990s and 2010s

Department	Extent of Reductions		Reduction Strategy		Workforce Impact	
	1990s	2010s	1990s	2010s	1990s	2010s
Army	Substantial	Substantial	Hiring freezes, voluntary separations, limited RIFs	Hiring freezes, voluntary separations, limited RIFs	Blue-collar, clerical; junior employees; minorities, women?	Nonprofessional employees; blacks, women
Air Force	Substantial	Small	Hiring freezes, voluntary separations, limited RIFs	Hiring freezes, voluntary separations, very limited RIFs	Blue-collar, clerical; junior employees; minorities, women?	Hispanics, women, persons with disabilities; junior employees
Navy	Substantial	Small	Hiring "frosts," voluntary separations, limited RIFs	Hiring "frosts," voluntary separations, negligible RIFs	Blue-collar, clerical; junior employees; minorities, women?	No significant impact expected

table indicates, the size of the current civilian drawdown is likely to be smaller than the previous one. That said, the Army will probably experience substantial reductions in the coming years, while the Air Force and Navy will face more limited cutbacks. Despite claims that they are becoming more analytic in their downsizing approaches, the services are using a similar strategy to reduce the civilian workforce today as they did in the 1990s. Specifically, the general strategy involves (1) implementing hiring freezes and offering voluntary incentives and early retirements; (2) instituting RIFs sparingly and only when necessary; and (3) avoiding furloughs if possible. Although neither drawdown appears to have significantly changed the overall composition of the civilian workforce (except in terms of aging), members of certain groups seem to have been disproportionately affected. In the 1990s, these groups included blue-collar and clerical workers, junior employees, and, possibly, minorities and women. Recent involuntary separations have somewhat disproportionately affected black and female civilians in the Army. For its part, the Air Force is concerned about a recent rise in mostly voluntary departures of Hispanic, female, and disabled civilians. The Navy does not expect that the reductions currently planned for its civilian workforce will have a disproportionate impact on women and minorities.

Methodology and Additional Results for Chapters Two to Four

This appendix offers additional details on the methods and results for Chapters Two through Four. Specifically, we provide the following:

- published-source review methodology
- interview methodology
- description of variables in the personnel files
- technical description of equation to decompose population change into accessions and separations
- technical description of modeling approach used to adjust CCRs for women and minorities
- additional CCR results.

All topics listed above apply to Chapters Two and Three. Only the first two topics apply to Chapter Four.

Published-Source Review

We began with a search for relevant published literature on past and recent drawdowns, as related to issues of demographic diversity. For our online search of Google Scholar, we looked for publications between June 1989 and June 2014. We combined terms related to a drawdown (e.g., "military downsizing," "military drawdown," and "reductions in force") with terms related to diversity (e.g., "diversity," "equal opportunity") and/or demographic categories (e.g., "women" and "minority"). Our initial hit rate was over 1,000 publications, although many of these were duplicates and did not meet our inclusion criteria. To be included in our review, publications had to be written in English, address demographic diversity, and meet minimum scholarly criteria (i.e., peer-reviewed or reviewed by master's/doctoral dissertation committee and involve a research study or a comprehensive review of scholarly literature). We therefore excluded opinion pieces, patents, book reviews, and other nonscholarly work. This search yielded only about 20 relevant publications.

To provide additional background on the drawdowns, we conducted another electronic search but cast the net more widely. We searched several electronic databases, including WorldCat, Web of Science, the Defense Technical Information Center (DTIC), GAO website, RAND's Online Catalog System (ROCS), and Air University Index to Military Periodicals. We also removed search terms related to diversity and instead combined terms relating to drawdowns (e.g., "drawdown," "downsizing," "reduction in force") with terms related to the U.S. military (e.g., "Department of Defense," "U.S. military," "Army," "Navy," "Marine Corps," "Air Force"). Moreover, we included fact-based news articles (i.e., not opinion pieces) in our search, as news articles often report details about drawdown programs, such as the type of program used and the size of cuts. The initial search yielded over 1,000 hits, but many were redundant. We identified about 100 publications (including news stories) with relevant details on drawdown policies, strategies, programs, and outcomes.

We supplemented our electronic database searches with snowball methodology. Specifically, we reviewed the reference lists of publications from our electronic search to identify other publications that might be relevant. We also consulted with colleagues knowledgeable about force drawdowns in the military and asked the experts we interviewed for recommendations on publications to read. We identified about 20 additional publications through snowball methods.

For Chapter Four, we performed another online search using Google to find news stories or details from websites posted by the services about recent drawdown programs. We focused on stories from 2011 to 2015. We tailored our search to each service, based on information provided by experts we interviewed about recent drawdown programs and tools. For example, Army experts discussed the Army's planned use of the Qualitative Service Program. We therefore searched for news stories or Army websites that discussed that program. This additional search was largely to support building the scenarios we used in Chapter Five; news stories provided some details about drawdown programs not available from our interviews.

Interviews

From fall 2013 to summer 2014, we conducted interviews with experts about past and recent drawdowns. The interviews were semistructured in nature, allowing the interviewer to ask follow-on questions not in the protocol if needed. Each interview lasted about an hour, was held in person or by phone, and involved at least one primary interviewer and one note-taker. Given the simple structure of the interviews, a systematic coding scheme was not used. The note-taker and interviewers reviewed the notes to identify themes and at least one other team member reviewed the themes.

Below, we provide a sample of the recruitment email sent to each participant and the semistructured interview protocol.

Recruitment Materials

Dear [Interviewee],

RAND's National Defense Research Institute is conducting a study for OSD's Office of Diversity Management and Equal Opportunity on the potential impact of force drawdowns on diversity within DoD's military and civilian workforces.

As part of this effort, we are gathering information from experts such as yourself on the current state of drawdown policies and practices across the services and OSD, as well as on past policies and practices, particularly during the last major drawdown in the 1990s.

I am hoping you might be the right person to speak with us about current and past drawdowns in relation to the [service component and population, e.g., Army Reserve]. If you do not think you are the right person to speak with us, would you be able to point us in the right direction? If possible, we would like to schedule an interview [in person/by phone] in the next couple of weeks. I anticipate the conversation would last about 45–60 minutes.

Thank you for your help.

Interview Protocol

We used background questions to determine whether to ask experts about past or current drawdowns and whether to ask about military or civilian reductions. Military reduction questions were framed for active-duty, Guard, or Reserve expertise. In most cases, we knew in advance what expertise the person would have. If someone expressed expertise across drawdown eras or personnel categories, we combined questions. For example, we might ask about goals during the 1990s military drawdown and then ask about the goals for current military drawdown. This made discussions flow more smoothly.

Background Questions

1. Can you briefly describe your current position?
2. Were you involved with drawdown policy and decisionmaking of the 1990s force drawdown? If so, how?
   [If "yes," frame questions for the 1990s drawdown.]
3. Have you been involved with recent drawdown policy and decisionmaking? If so, how?
   [If "yes," frame questions for recent drawdown(s).]
4. For the drawdown eras with which you are familiar, does your expertise extend to the civilian defense workforce?
   [If "yes," frame questions for civilian reductions.]

Drawdown Questions

1. What [were/are] the [name of service/DoD] goals for the [1990s/current] drawdown?
   – How [did/do] those goals differ [from the other/across the] services?

- [If civilian drawdown expert:] [Did/Do] civilian reduction goals differ from those for military personnel? If so, how?
2. What strategies [did/do] the [name of service/services] use for military force reductions? [*If expert on civilian reductions, replace "military force" with "civilian."*]
    - How [did/do] these strategies differ [from other/across the] services?
    - [If civilian drawdown expert:] [Did/Do] civilian reduction strategies differ from those for military personnel? If so, how?
    - [Did/Do] these strategies influence demographic diversity within the [name of service/services]? If so, how?
3. [Was/Is] diversity taken into account by the [name of service/services] for the [1990s/current] drawdown? If so, how?
    - [Probe:] [Was/Is] diversity taken into account explicitly (e.g., used in force-shaping models)? Implicitly (e.g., targeting factors that correlate with demographics without formally including diversity as a factor)?
4. In general, what [were/are] the most influential laws and policies affecting drawdown decisions [in the 1990s/for the current drawdown]?
    - Do any of them relate to diversity? If so, in what way(s)?
5. Can you offer any documents related to past or current drawdowns for us to review?
6. Are there other experts on past or current drawdowns that you recommend we contact?
7. Are there additional topics related to diversity during drawdowns that you wish to discuss? Any other questions?

Variables in Personnel Files

In this section, we review the DMDC active-duty master file data used in the study. In the next two sections, we provide more details about our data analysis, namely the decomposition of population change into accessions and separations and our retention analyses.

Non-Demographic Variables

Most of the variables used in our analyses were not significantly modi-fied except to check for suspicious data patterns (e.g., high levels of missing data). In Table C.1, we provide brief descriptions of the vari-ables. The variable coding scheme used for the CCR models is described under "Retention Analyses" later in the appendix.

Demographic Variables

DMDC offers three variables to capture the self-reported racial and ethnic identities of military personnel. One variable captures a per-son's race, another captures his/her Hispanic ethnic status, and a third reflects the various ethnic groups to which the individual belongs. We used two of the variables, race and Hispanic status, to create five racial/ethnic categories that generally align with those used by the U.S. Census Bureau: non-Hispanic white, non-Hispanic black, Hispanic, non-Hispanic Asian, and other (includes non-Hispanic American Indi-ans, non-Hispanic Alaskan Natives, and non-Hispanic individuals who select more than one race). The one exception from the Census Bureau categorization is that our non-Hispanic Asian category includes both Asian and Hawaiian/Pacific Islander categories. We combined these two categories because the race categories changed during the period of our analysis and because of a coding error in the DMDC master data-set.[1] Although most of our analyses focused on comparisons among non-Hispanic whites, non-Hispanic blacks, and Hispanics, we some-times used a majority-minority comparison whereby non-Hispanic whites were coded as "majority" and the combined set of the remain-ing racial/ethnic groups (except for those coded as "unknown" or oth-erwise missing race or ethnicity information) were coded as "minority."

For both gender and race/ethnicity, we reviewed individuals' records to look for changes over time. For the most part, individuals who reported being in a given category retained that category through-

---

[1]   Prior to 1995, the available codes were "white," "black," and "other." In 1995, Ameri-can Indian/Alaskan Native and Asian were added. In 2003, Hawaiian/Pacific Islander and several multirace categories (e.g., white and black) were made available. For FY 2002, the DMDC data include an error whereby Asian and Hawaiian/Pacific Islander were lumped together into a single category.

Table C.1
DMDC Variables Used for Historical Drawdown Analyses

Variable	Description
AFQT category	Categories constructed from AFQT variable in DMDC file. Categories correspond to the following AFQT score ranges: Category I: 93–100 Category II: 65–92 Category IIIA: 50–64 Category IIIB: 31–49 Category IV and V: 0–30
Education category— enlisted	Categories constructed from DMDC education variable. Categories correspond to following DMDC education codes: <u>Less than high school diploma/alternative credential</u> = "Non–high school graduate," "Secondary school credential near completion," "Test-based equivalency diploma," "Occupational program certificate," "Correspondence school diploma," and "GED Certificate, ARNG Challenge Program." <u>High school diploma or equivalent</u> = "Attending high school, junior or less," "Attending high school, senior," "High school certificate of attendance," "Home study diploma," "Adult education diploma," "High school diploma," "Completed one semester of college, no high school diploma." <u>Some college</u> = "1 year of college certificate of equivalence," "1–2 years of college, no degree," "Associate degree," and "Professional nursing diploma." <u>Baccalaureate degree or higher</u> = "Baccalaureate degree," "Master's degree," "Post master's degree," "First professional degree," "Doctorate degree," and "Post doctorate." <u>Unknown/missing</u> = personnel with "unknown" education codes or without any education information.
Education category— officer	Categories constructed from DMDC education variable. The same education categories from the DMDC education variable were used with one exception: Baccalaureate degree was differentiated from postbaccalaureate degrees.
Fiscal year (FY)	DMDC variable that provides the year (as of September) for each record.

Table C.1—Continued

Variable	Description
Officer accession source	Categories constructed from DMDC variable on "military accession program." Constructed categories include the following DMDC categories: • <u>Military academy</u> = U.S. Military Academy, U.S. Naval Academy, U.S. Air Force Academy, U.S. Coast Guard Academy, U.S. Merchant Marine Academy, Air National Guard Academy of Military Sciences. • <u>Reserve Officers' Training Corps (ROTC)</u> = ROTC/Naval ROTC [NROTC] scholarship program and ROTC/NROTC nonscholarship program • <u>Officer Candidate School/Officer Training School</u> = "OCS, AOCS, OTS, or PLC" and "National Guard State OCS" • <u>Other</u> = any code other than those in previous categories, missing, or "unknown" codes. Examples: "Direct appointment authority" and "Aviation training program other than OCS, AOCS, OTS, or PLC." • <u>Unknown/missing</u> = officers with "unknown" codes or without any commissioning source information
Occupational category— enlisted	DoD occupational categories for enlisted personnel include • craftsworkers • communications and intelligence specialists • electrical/mechanical equipment repairers • electronic equipment repairers • functional support and administration • health care specialists • infantry, gun crews, and seamanship specialists • nonoccupational (e.g., trainees) • other technical and allied specialists • service and supply handlers. For tactical vs. nontactical comparisons, tactical occupations were equated with the category "Infantry, Gun Crews, and Seamanship Specialists."
Occupational category— officer	DoD occupational categories for officers include • administrators • engineering and maintenance officers • general officers and executives, N.E.C. • health care officers • intelligence officers • nonoccupational (e.g., cadets) • scientists and professionals • supply, procurement and allied officers • tactical operations officers For tactical vs. nontactical comparisons, tactical occupations were equated with "Tactical Operations Officers."
RAND ID	Variable that provides a code for each individual in the file.

Table C.1—Continued

Variable	Description
Rank	DMDC variable where E-1 through E-9 corresponds to enlisted ranks; W-1 through W-5 to warrant officer ranks; and O-1 through O-10 corresponds to commissioned officer ranks. Warrant officer ranks were not used in study because not all services have warrant officers and because of small population-size concerns.
Rank group (corps)	To group data into enlisted and commissioned officer corps, the rank variable was used. Any record with an enlisted rank was classified as enlisted corps, and any record with an officer rank was classified as officer corps.
Service	DMDC variable that specifies service: Army (A), Navy (N), Marine Corps (M), and Air Force (F).
Time in grade (TIG)	Constructed by subtracting the date of rank (e.g., date when system indicates person went from E-4 to E-5) from the date of the record. Values reflect number of months at the grade (e.g., 12 months in E-5).
Years of service (YOS)	Constructed by subtracting the individual's active-duty federal military start date from the date of the record. Values are rounded down to whole numbers to reflect years of service. YOS carries over from previous service periods for personnel with breaks in service.

out their time in service. However, a small minority of individuals changed gender categories or racial/ethnic categories. For gender, we used the most recent category to backfill earlier years. Less than 0.2 percent of the records from FY 1989–2012 resulted in gender switches (female to male or male to female). In cases where the most recent record was missing gender, the last record for which gender was available was used to backfill and forward fill. For example, an individual with the following pattern over four years—Year 1 = Male, Year 2 = missing, Year 3 = Female, and Year 4 = missing—was classified as "Female" for all four years.

For race/ethnicity, we used a more complex data-cleaning strategy because we expected that some personnel change their racial/ethnic identities over time, especially as more racial/ethnic categories became available over time. Our first step was to make the race codes consistent over time before merging with the Hispanic ethnicity codes. We used the following decision rules to reclassify race codes for individuals with multiple records:

1. Multirace codes trumped all other race codes, so if an individual had one of the multirace codes in any year, then all of the person's records were designated "Multirace."
2. American Indian/Alaskan Native trumped all other race categories except Multirace.
3. If an individual had two distinct minority race categories (e.g., Black and Asian), the individual's records were coded as Multirace.
4. If an individual had a combination of one minority code and the majority (White) code, that person's records were coded as the minority code.
5. Any race code trumps missing or unknown codes (e.g., two of three years with Black codes and one year with missing code resulted in missing code changed to Black).

Next, we reclassified the Hispanic ethnicity codes (i.e., Hispanic, non-Hispanic, Unknown, and missing) using the following rule set:

1. If an individual listed ethnicity as Hispanic at any point, the other records for that person were categorized as "Hispanic."
2. Any code designating a known Hispanic status (i.e., Hispanic or non-Hispanic) trumps missing or unknown codes (e.g., two of three years with the non-Hispanic code and one year with the Unknown code resulted in Unknown code being changed to non-Hispanic).

Once we adjusted the race and Hispanic variables, we combined them such that individuals with Hispanic status were classified as "Hispanic" and individuals with non-Hispanic status were classified according to their race codes (e.g., non-Hispanic black).

Decomposition of Population Change into Accessions and Separations

Methodology

We developed an equation that would decompose the annual change in female or minority group representation in the force into inflows (accessions) and outflows (separations). The equation for change in female representation has the following form:

$$\frac{f_t}{p_t} - \frac{f_{t-1}}{p_{t-1}} = \frac{m_{t-1} f_{t-1}}{p_{t-1} p_t} \left(\frac{a_f}{f_{t-1}} - \frac{a_m}{m_{t-1}} \right) + \frac{m_{t-1} f_{t-1}}{p_{t-1} p_t} \left(\frac{s_m}{m_{t-1}} - \frac{s_f}{f_{t-1}} \right),$$

where f and m stand for the number of women and men in the population, respectively; p is the total population (for a given service-corps combination, e.g., Army enlisted); t represents the fiscal year; a_f is number of female accessions entering the population at the beginning of year t; a_m is number of male accessions entering the population at the beginning of year t; s_f is number of women separating from the population at the end of year $t-1$; s_m is number of men separating from the population at the end of year $t-1$.

The equation for non-Hispanic blacks replaces f with b and m with nb for all demographic groups except non-Hispanic blacks. A similar logic applies to Hispanic representation change.

The equation controls for changes not only in the female population but also in the population as a whole. The first part of the equation (before the plus sign) represents the change in accessions, with positive values reflecting a higher female representation among accessions relative to the rest of the population (i.e., males). The second half of the equation (after the plus sign) represents change in separations, with positive values reflecting lower female representation among separators relative to the rest of the population. In short, positive values reflect increases in female representation from one year to the next, whereas negative values reflect decreases in female representation from one year to the next. The same logic applies for our non-Hispanic black and Hispanic results, although the comparison groups are all other racial/

ethnic groups, which includes both non-Hispanic white and other minority groups.

Additional Results

Next, we provide the total population sizes by service and corps and results on demographic changes for all demographic groups, services, and corps for the 1990s reductions and 2000s reductions. To conserve space, the demographic change results are presented in tables instead of charts. Table C.2 provides the population sizes for the 1990s drawdown and postdrawdown years. The 1990s demographic change results are in Tables C.3 to C.6. The population sizes for the 2000s drawdown and postdrawdown periods for the Navy and Air Force are in Table C.7. The main 2000s demographic change results are in Tables C.8 to C.9. Table C.10 shows additional decomposition results from a gender-by-race/ethnicity breakout analysis for Air Force enlisted. We show results to two decimal places in Table C.10 due to the small sizes in the groups that result in smaller changes in percentage points.

For each demographic group, we provide the two components of change due to accessions and change due to separations. To facilitate the interpretation of results, we present the components of change in terms of percentage points. Total change is the sum of the two components. We bold values less than zero to identify trends more easily.

We offer an example to aid readers in interpreting values in the tables. In Table C.3, the first data column shows values for changes in the Army enlisted female population between 1989 and 1990, as well as the total population size from 1989. The 0.2 "Total" value reflects the total change in female representation from 1989 to 1990. Specifically, 10.9 percent of the Army enlisted population in 1989 was women, and 11.1 percent of the Army enlisted population in 1990 was women. The difference between 11.1 percent and 10.9 percent is 0.2 percentage points. Multiply 0.2 percent times the 1989 population of 654,299 (from Table C.2) for a value of approximately 1,309, which can be interpreted as the equivalent change in the female population had the male population not significantly changed in size between 1989 and 1990.

The 0.2 percentage-point change is composed of a 0.3 percentage-point change due to accession trends and a −0.1 percentage-point change due to separation trends. The accession and separation percentage-point changes can be multiplied by the 1989 population size, resulting in values of 1,963 for accessions and −654 for separations. Had the Army enlisted force not lost or gained any men between 1989 and 1990, the equivalent female demographics would have been a gain of 1,963 women and a loss of 654 women, for a total gain of 1,309 women between 1989 and 1990.

Table C.2
Population Sizes by Service and Corps During and After 1990s Drawdown

Fiscal Year	Army Enlisted	Army Officer	Navy Enlisted	Navy Officer	Marine Corps Enlisted	Marine Corps Officer	Air Force Enlisted	Air Force Officer
1989	654,299	91,713	512,914	69,216	176,586	18,455	462,682	103,679
1990	620,192	89,030	499,681	69,174	175,902	18,055	430,730	100,035
1991	600,860	87,580	491,804	67,790	173,758	17,757	409,271	96,586
1992	507,561	79,287	465,183	66,090	162,711	17,016	375,332	90,360
1993	479,179	74,695	436,918	63,522	157,746	16,484	355,988	84,067
1994	449,744	72,273	400,096	58,929	155,465	15,930	341,108	80,985
1995	416,014	70,249	369,784	56,205	157,278	15,720	320,632	79,680
1996	399,700	68,574	353,016	55,417	158,045	16,012	308,328	76,787
1997	399,366	67,357	324,168	54,095	154,950	15,977	299,037	74,474
1998	400,538	66,788	319,570	52,878	154,110	15,773	291,296	71,806
1999	392,424	65,223	308,436	51,875	152,865	15,982	282,537	70,080
2000	395,732	65,024	311,869	50,645	153,677	15,957	281,108	68,979
2001	393,388	64,403	316,562	51,523	154,129	16,153	276,568	67,666

Table C.3
Demographic Changes During and After 1990s Drawdown: Army

Demographic Group	1990	1991	1992	1993	1994	1995	1996	1997	1998	1999	2000	2001
Enlisted												
Female												
Accessions	0.3	0.2	0.5	0.3	0.5	0.5	0.8	0.7	0.3	0.4	0.8	0.5
Separations	−0.1	−0.3	0.3	0.2	0.0	0.1	0.0	0.0	−0.2	−0.4	−0.5	−0.3
Total	0.2	−0.1	0.8	0.5	0.5	0.6	0.8	0.7	0.1	0.1	0.3	0.2
Black												
Accessions	−0.9	−1.4	−1.6	−1.6	−1.2	−1.0	−1.1	−1.1	−1.0	−0.9	−1.1	−1.1
Separations	1.8	1.2	1.3	0.9	0.8	1.1	1.0	0.9	0.7	0.8	0.8	0.9
Total	0.9	−0.2	−0.3	−0.8	−0.4	0.1	0.0	−0.3	−0.3	−0.1	−0.3	−0.2
Hispanic												
Accessions	0.0	−0.1	0.0	0.0	0.0	0.2	0.3	0.4	0.3	0.3	0.3	0.2
Separations	0.3	0.3	0.5	0.4	0.4	0.4	0.3	0.3	0.3	0.3	0.3	0.2
Total	0.3	0.2	0.5	0.4	0.4	0.5	0.7	0.8	0.6	0.6	0.5	0.4
Officer												
Female												
Accessions	0.5	0.5	0.4	0.4	0.4	0.4	0.1	0.4	0.4	0.6	0.5	0.5
Separations	−0.1	−0.3	0.0	−0.1	−0.3	−0.3	−0.2	−0.5	−0.3	−0.2	0.0	0.0
Total	0.4	0.2	0.4	0.3	0.1	0.1	−0.1	−0.1	0.2	0.4	0.5	0.5
Black												
Accessions	0.1	0.0	0.0	−0.2	0.0	−0.1	−0.1	−0.1	0.0	0.0	0.1	0.2
Separations	0.2	0.1	0.2	−0.1	0.1	0.1	0.0	−0.1	0.2	0.1	0.2	0.2
Total	0.3	0.2	0.2	−0.2	0.0	0.0	−0.1	−0.2	0.1	0.1	0.3	0.4

Table C.3—Continued

Demographic Group	1990	1991	1992	1993	1994	1995	1996	1997	1998	1999	2000	2001
					Officer							
Hispanic												
Accessions	0.1	0.1	0.1	0.0	0.1	0.1	0.1	0.1	0.1	0.1	0.2	0.1
Separations	0.1	0.1	0.2	0.1	0.1	0.1	0.1	0.0	0.1	0.1	0.1	0.1
Total	0.2	0.2	0.3	0.1	0.2	0.2	0.2	0.1	0.2	0.2	0.3	0.2

SOURCE: Analysis of DMDC data on active-duty Army personnel (FY 1990–2001)

1990s Drawdown

Table C.4
Demographic Changes During and After 1990s Drawdown: Navy

Demographic Group	1990	1991	1992	1993	1994	1995	1996	1997	1998	1999	2000	2001
					Enlisted							
Female												
Accessions	0.2	0.0	0.5	0.3	0.6	0.9	0.3	0.3	0.8	0.6	0.7	0.6
Separations	0.0	0.0	−0.1	0.0	−0.1	0.1	−0.1	−0.2	−0.3	−0.3	−0.2	−0.1
Total	0.2	−0.1	0.4	0.3	0.5	10.0	0.3	0.1	0.5	0.3	0.4	0.4
Black												
Accessions	0.3	−0.3	−0.2	−0.3	−0.1	0.1	0.1	0.0	0.1	0.0	0.1	0.1
Separations	0.5	0.3	0.3	0.3	0.4	0.6	0.4	0.5	0.2	0.3	0.2	0.3
Total	0.8	0.0	0.1	0.0	0.3	0.7	0.5	0.5	0.3	0.4	0.3	0.4
Hispanic												
Accessions	0.4	0.4	0.4	0.2	0.1	0.5	0.6	0.5	0.5	0.4	0.5	0.5
Separations	0.1	0.0	0.1	0.0	0.0	0.0	0.1	0.0	0.1	0.1	0.1	0.2
Total	0.5	0.5	0.5	0.3	0.0	0.5	0.6	0.5	0.6	0.6	0.5	0.6

Table C.4—Continued

Demographic Group	1990	1991	1992	1993	1994	1995	1996	1997	1998	1999	2000	2001
						Officer						
Female												
Accessions	0.5	0.4	0.6	0.3	0.2	0.3	0.2	0.2	0.2	0.3	0.4	0.4
Separations	0.0	0.0	0.1	0.1	0.3	0.2	0.0	0.0	0.0	−0.1	−0.1	−0.2
Total	0.5	0.4	0.7	0.4	0.5	0.5	0.2	0.2	0.3	0.2	0.3	0.2
Black												
Accessions	0.2	0.1	0.1	0.2	0.2	0.2	0.1	0.2	0.2	0.1	0.1	0.1
Separations	0.0	0.1	0.1	0.1	0.1	0.2	0.1	0.1	0.1	0.2	0.1	0.1
Total	0.3	0.2	0.2	0.2	0.4	0.4	0.3	0.3	0.3	0.3	0.2	0.2
Hispanic												
Accessions	0.2	0.1	0.1	0.2	0.2	0.2	0.2	0.2	0.2	1.2	0.1	0.2
Separations	0.0	0.1	0.1	0.1	0.1	0.1	0.1	0.2	0.1	0.1	0.3	0.2
Total	0.2	0.2	0.2	0.2	0.3	0.4	0.2	0.4	0.3	1.4	0.4	0.4

SOURCE: Analysis of DMDC data on active-duty Navy personnel (FY 1990–2001). The 1999 results for Hispanic officers have a very high value for accessions. We could not identify a reason this year experienced such an increase in accession records. Please regard the result with caution.

Table C.5
Demographic Changes During and After 1990s Drawdown: Marine Corps

Demographic Group	1990	1991	1992	1993	1994	1995	1996	1997	1998	1999	2000	2001
						Enlisted						
Female												
Accessions	0.0	0.0	0.0	−0.1	0.0	0.2	0.2	0.4	0.3	0.2	0.1	0.1
Separations	−0.1	−0.1	−0.1	−0.1	0.0	0.0	0.0	0.2	0.0	0.1	0.0	0.0
Total	−0.1	−0.2	−0.1	−0.2	0.0	0.2	0.2	0.5	0.3	0.2	0.1	0.1

Table C.5—Continued

Demographic Group	1990	1991	1992	1993	1994	1995	1996	1997	1998	1999	2000	2001
					Enlisted							
Black												
Accessions	−0.7	−1.1	−1.3	−1.4	−1.0	−0.7	−0.6	−0.5	−0.7	−0.7	−0.7	−0.7
Separations	0.6	0.5	0.3	0.4	0.3	0.6	0.4	0.5	0.5	0.5	0.3	0.3
Total	−0.1	−0.6	−1.0	−1.0	−0.7	−0.1	−0.2	0.0	−0.2	−0.1	−0.3	−0.4
Hispanic												
Accessions	0.2	0.1	0.1	0.3	0.2	0.6	0.6	0.4	0.4	0.4	0.4	0.4
Separations	0.3	0.3	0.3	0.4	0.3	0.4	0.4	0.4	0.4	0.3	0.3	0.4
Total	0.5	0.4	0.4	0.7	0.6	1.0	0.9	0.8	0.8	0.7	0.7	0.7
					Officer							
Female												
Accessions	0.0	0.1	−0.1	0.2	0.2	0.3	0.4	0.3	0.5	0.3	0.3	0.4
Separations	−0.1	0.0	−0.1	−0.1	−0.1	0.0	0.0	0.0	−0.1	0.0	−0.1	−0.1
Total	−0.1	0.1	−0.2	0.1	0.1	0.3	0.3	0.3	0.4	0.3	0.2	0.2
Black												
Accessions	0.0	0.0	0.1	0.1	0.2	0.2	0.3	0.3	0.3	0.1	−0.1	−0.1
Separations	−0.1	0.0	−0.1	−0.1	−0.1	0.0	0.0	0.1	0.1	0.0	0.1	0.0
Total	−0.1	0.0	0.0	0.0	0.2	0.2	0.3	0.4	0.4	0.1	0.0	−0.1
Hispanic												
Accessions	0.1	0.1	0.1	0.1	0.2	0.4	0.3	0.2	0.4	0.4	0.2	0.3
Separations	0.1	0.1	0.1	0.0	0.1	0.0	0.1	0.2	0.1	0.1	0.1	0.1
Total	0.2	0.2	0.2	0.1	0.2	0.3	0.4	0.4	0.5	0.5	0.3	0.4

SOURCE: Analysis of DMDC data on active-duty Marine Corps personnel (FY 1990–2001).

Table C.6
Demographic Changes During and After 1990s Drawdown: Air Force

Demographic Group	1990	1991	1992	1993	1994	1995	1996	1997	1998	1999	2000	2001
Enlisted												
Female												
Accessions	0.4	0.4	0.6	0.5	0.6	0.7	0.8	1.0	0.8	0.8	0.8	0.5
Separations	−0.1	−0.1	−0.1	−0.2	−0.1	−0.1	−0.2	−0.1	−0.1	−0.3	−0.3	−0.2
Total	0.3	0.3	0.5	0.4	0.5	0.6	0.7	0.9	0.6	0.5	0.4	0.3
Black												
Accessions	−0.4	−0.5	−0.5	−0.4	−0.2	−0.2	−0.1	0.0	0.1	0.0	0.1	**−0.1**
Separations	0.6	0.2	0.3	0.1	0.2	0.2	0.4	0.3	0.3	0.2	0.3	0.2
Total	0.3	**−0.3**	**−0.2**	**−0.3**	**−0.1**	0.1	0.2	0.3	0.3	0.3	0.3	0.1
Hispanic												
Accessions	0.0	0.0	0.0	0.0	0.1	0.2	0.2	0.2	0.3	0.3	0.3	0.2
Separations	0.1	0.0	0.1	0.1	0.1	0.1	0.1	0.1	0.1	0.1	0.1	0.2
Total	0.1	0.0	0.1	0.1	0.2	0.3	0.3	0.4	0.4	0.4	0.4	0.4
Officer												
Female												
Accessions	0.5	0.5	0.4	0.4	0.5	0.4	0.6	0.5	0.4	0.5	0.4	0.4
Separations	**−0.1**	0.0	**−0.2**	0.1	0.1	**−0.3**	**−0.2**	**−0.1**	**−0.1**	**−0.3**	**−0.1**	**−0.2**
Total	0.4	0.5	0.2	0.5	0.7	0.2	0.4	0.4	0.4	0.2	0.3	0.2
Black												
Accessions	0.0	0.0	0.0	0.1	0.0	0.0	0.1	0.1	0.0	0.1	0.1	0.1
Separations	0.1	0.1	0.1	**−0.1**	0.0	0.0	0.0	0.1	0.1	0.1	0.1	0.0
Total	0.1	0.1	0.0	**−0.1**	0.0	0.0	0.2	0.2	0.1	0.1	0.2	0.1

Table C.6—Continued

Demographic Group	1990	1991	1992	1993	1994	1995	1996	1997	1998	1999	2000	2001
Hispanic												
Accessions	0.0	0.0	0.0	0.0	0.0	0.1	0.1	0.1	0.0	0.1	0.1	0.1
Separations	0.0	0.0	0.0	0.0	0.1	0.0	0.1	0.1	0.1	0.1	0.1	0.1
Total	0.0	0.0	0.1	0.0	0.1	0.1	0.1	0.2	0.1	0.2	0.3	0.2

SOURCE: Analysis of DMDC data on active-duty Air Force personnel (FY 1990–2001).

2000s Drawdown

We continued the analyses from the 1990s analysis, which involved years before the 2000s drawdowns. We modified the titles for Tables C.8 and C.9 to reflect this change. To supplement our analysis of Air Force enlisted gender trends in Chapter Two, we provide additional results broken out by race/ethnicity and gender categories in Table C.10. Because of the smaller group sizes, we rounded the results to two decimal places instead of one.

Retention Analyses

Methodology

A logistic regression model describes the relationship between a binary (0, 1) outcome variable and one or more predictor variables (Agresti, 1996). The binary response variable is measured in terms of log odds, or logits, which are the log of the odds of "success" (i.e., the odds of the outcome occurring). In the analyses in Chapters Two and Three, the odds of "success" refer to the probability of personnel continuing on active duty in the given service from year t to year $t + 1$ over the probability of not continuing on active duty in the given service from year t to year $t + 1$.

Like other generalized linear models, logistic regression models can handle multiple predictors that can be qualitative (categorical) or quantitative. In our analyses to adjust CCRs, we used several predictors that represent work-relevant characteristics that may differ between

Table C.7
Population Sizes by Service and Corps During and After Mid-2000s Drawdowns

	Navy		Air Force	
Fiscal Year	Enlisted	Officer	Enlisted	Officer
2000	311,869	50,645	281,108	68,979
2001	316,562	51,523	276,568	67,666
2002	323,902	52,750	291,360	71,422
2003	321,357	53,205	296,030	73,600
2004	312,722	52,609	297,089	74,104
2005	304,309	51,155	275,764	73,216
2006	292,669	50,318	273,705	70,510
2007	280,489	49,664	261,025	65,687
2008	275,001	49,637	257,770	64,757
2009	271,721	50,364	263,170	65,459
2010	270,424	51,036	263,214	66,173
2011	266,754	51,584	262,674	65,442

individuals. The predictors are as follows: YOS, gender, race/ethnicity, AFQT (enlisted only), accession source (officer only), education level, rank category, TIG, and period (FY). We treated YOS (0 and up) and AFQT (0–99) as continuous predictor variables. We treated the following variables as categorical predictors:

- **Accession source (officer):** Direct appointment or other (0), Officer Candidate School/Officer Training School/Platoon Leaders Class (1), ROTC (2), Military academy (3)
- **Education level (enlisted):** Unknown (0), Less than high school diploma/alternative credential (1), High school diploma or equivalent (2), Some college (3), Baccalaureate degree or higher (4)

Table C.8
Demographic Changes During and After Mid-2000s Drawdown: Navy

Demographic Group	2001	2002	2003	2004	2005	2006	2007	2008	2009	2010	2011
Enlisted											
Female											
Accessions	0.6	0.3	0.2	0.2	0.2	0.4	0.4	0.5	0.7	0.7	0.7
Separations	-0.1	-0.1	-0.1	-0.2	-0.3	-0.3	-0.2	-0.2	-0.1	-0.2	-0.2
Total	0.4	0.3	0.1	0.0	-0.1	0.1	0.2	0.3	0.5	0.5	0.5
Black											
Accessions	0.1	-0.4	-0.2	-0.2	-0.4	-0.6	-0.6	-0.5	-0.5	-0.6	-0.4
Separations	0.3	0.3	0.1	0.1	0.1	0.0	0.1	0.1	0.0	0.0	-0.1
Total	0.4	-0.1	-0.1	-0.1	-0.3	-0.6	-0.5	-0.3	-0.5	-0.6	-0.5
Hispanic											
Accessions	0.5	0.3	0.2	0.3	0.5	0.6	0.8	0.9	0.4	0.4	0.1
Separations	0.2	0.1	0.2	0.3	0.4	0.6	0.1	0.2	0.2	0.2	0.2
Total	0.6	0.5	0.4	0.6	0.9	0.0	0.9	0.0	0.6	0.6	0.3
Officer											
Female											
Accessions	0.4	0.3	0.3	0.3	0.3	0.3	0.4	0.5	0.5	0.5	0.6
Separations	-0.2	-0.2	-0.3	-0.3	-0.5	-0.4	-0.3	-0.2	-0.2	-0.2	-0.2
Total	0.2	0.1	0.0	-0.1	-0.2	-0.1	0.1	0.3	0.6	0.3	0.4

Table C.8—Continued

Demographic Group	2001	2002	2003	2004	2005	2006	2007	2008	2009	2010	2011
					Officer						
Black											
Accessions	0.1	0.0	0.1	0.0	0.0	0.1	-0.1	0.1	0.1	0.1	0.0
Separations	0.1	0.1	0.1	0.1	0.1	0.1	0.1	0.0	0.0	0.0	0.0
Total	0.2	0.1	0.2	0.0	0.1	0.1	0.0	0.2	0.1	0.1	0.0
Hispanic											
Accessions	0.2	0.1	0.0	-0.1	0.0	0.0	0.0	0.0	0.1	0.0	0.0
Separations	0.2	0.2	0.2	0.0	0.0	0.1	0.0	0.0	0.0	0.1	0.1
Total	0.4	0.3	0.2	-0.1	0.0	0.0	0.0	0.1	0.1	0.2	0.1

SOURCE: Analysis of DMDC data on active-duty Navy personnel (FY 2001–2011).

Table C.9
Demographic Changes During and After Mid-2000s Drawdown: Air Force

Demographic Group	2001	2002	2003	2004	2005	2006	2007	2008	2009	2010	2011
Enlisted											
Female											
Accessions	0.5	0.5	0.4	0.1	0.1	0.3	0.2	0.1	0.0	−0.1	−0.1
Separations	−0.2	−0.2	−0.2	−0.2	−0.2	−0.2	−0.3	−0.2	−0.2	−0.3	−0.1
Total	0.3	0.3	0.1	0.0	0.0	0.2	−0.1	−0.1	−0.2	−0.4	−0.2
Black											
Accessions	−0.1	−0.4	−0.5	−0.4	−0.2	−0.3	−0.1	0.0	−0.1	−0.1	−0.2
Separations	0.2	0.0	0.0	0.1	0.1	0.0	0.1	0.1	0.0	0.0	−0.1
Total	0.1	−0.4	−0.4	−0.4	−0.1	−0.3	−0.1	0.1	0.0	−0.1	−0.2
Hispanic											
Accessions	0.2	0.2	−0.2	−0.4	−0.2	−0.3	−0.3	−0.3	−0.4	−0.4	−0.5
Separations	0.2	0.1	0.1	0.1	0.2	0.1	0.1	0.1	0.1	0.1	0.1
Total	0.4	0.3	−0.1	−0.2	−0.1	−0.3	−0.2	−0.2	−0.4	−0.3	−0.3
Officer											
Female											
Accessions	0.4	0.6	0.5	0.5	0.5	0.3	0.4	0.6	0.4	0.5	0.4
Separations	−0.2	−0.1	−0.1	−0.3	−0.4	−0.5	−0.6	−0.3	−0.3	−0.3	−0.3
Total	0.2	0.5	0.3	0.1	0.1	−0.2	−0.2	0.3	0.2	0.2	0.1
Black											
Accessions	0.1	0.2	0.0	0.0	0.0	0.1	0.1	0.0	0.1	0.0	0.0
Separations	0.0	0.0	0.0	0.0	0.0	0.1	0.1	0.1	0.0	0.0	0.1
Total	0.1	0.2	0.0	0.1	0.1	0.2	0.2	0.1	0.1	0.0	0.1

Table C.9—Continued

Demographic Group	2001	2002	2003	2004	2005	2006	2007	2008	2009	2010	2011
Officer											
Hispanic											
Accessions	0.1	0.2	0.1	0.0	0.1	0.1	0.1	0.1	0.1	0.1	0.1
Separations	0.1	0.1	0.1	0.1	0.1	0.0	0.0	0.0	0.0	0.0	0.0
Total	0.2	0.3	0.2	0.2	0.0	−0.1	−0.1	−0.1	−0.1	−0.1	−0.1

SOURCE: Analysis of DMDC data on active-duty Air Force personnel (FY 2001–2011).

Table C.10
Additional Decomposition Results for Air Force Enlisted Population During the 2000s: Gender-by-Race/Ethnicity Groups

Demographic Group	2001	2002	2003	2004	2005	2006	2007	2008	2009	2010	2011
Non-Hispanic white men											
Accessions	−0.63	−0.38	−0.02	0.24	0.04	−0.04	−0.01	0.07	0.20	0.36	0.25
Separations	−0.18	0.02	0.00	−0.03	−0.15	0.01	0.01	−0.06	0.04	0.07	0.00
Total	−0.81	−0.36	−0.02	0.21	−0.10	−0.02	0.01	0.01	0.24	0.43	0.25
Non-Hispanic white women											
Accessions	0.23	0.33	0.44	0.25	0.18	0.37	0.27	0.19	0.17	0.09	0.05
Separations	−0.31	−0.24	−0.28	−0.27	−0.29	−0.25	−0.36	−0.27	−0.26	−0.31	−0.20
Total	−0.09	0.10	0.16	−0.02	−0.11	0.12	−0.09	−0.07	−0.09	−0.23	−0.15
Non-Hispanic black men											
Accessions	−0.18	−0.38	−0.35	−0.28	−0.14	−0.21	−0.07	0.00	0.00	0.00	−0.03
Separations	0.15	0.03	0.00	−0.01	0.08	0.00	0.04	0.07	0.02	0.02	−0.10
Total	−0.04	−0.35	−0.35	−0.30	−0.06	−0.22	−0.04	0.07	0.02	0.02	−0.13

Table C.10—Continued

Demographic Group	2001	2002	2003	2004	2005	2006	2007	2008	2009	2010	2011
Non-Hispanic black women											
Accessions	0.06	−0.04	−0.12	−0.14	−0.08	−0.07	−0.05	−0.02	−0.08	−0.14	−0.13
Separations	0.06	0.02	0.02	0.07	0.06	0.03	0.02	0.04	0.03	0.02	0.03
Total	0.12	−0.02	−0.10	−0.07	−0.02	−0.04	−0.03	0.02	−0.05	−0.12	−0.10
Hispanic men											
Accessions	0.10	0.14	−0.17	−0.30	−0.18	−0.28	−0.27	−0.26	−0.34	−0.34	−0.36
Separations	0.14	0.10	0.10	0.11	0.12	0.06	0.10	0.09	0.08	0.12	0.09
Total	0.25	0.24	−0.08	−0.18	−0.06	−0.22	−0.17	−0.18	−0.27	−0.22	−0.28
Hispanic women											
Accessions	0.08	0.10	−0.02	−0.08	−0.05	−0.06	−0.07	−0.09	−0.10	−0.10	−0.11
Separations	0.02	0.01	0.01	0.03	0.04	0.02	0.02	0.02	0.01	0.02	0.03
Total	0.11	0.12	−0.01	−0.05	−0.01	−0.05	−0.05	−0.06	−0.09	−0.08	−0.07

SOURCE: Analysis of DMDC data on active-duty Air Force personnel (FY 2001–2011).

- **Education level (officer):** Unknown (0), High school diploma or equivalent (1), Some college (2), Baccalaureate degree (3), Post-baccalaureate degree (4)
- **Occupational category (officer):** nontactical operations (0), tactical operations (1)
- **Occupational category (enlisted):** nontactical operations (0), Infantry, gun crews, and seamanship specialists (1)
- **Rank category (enlisted):** Up to E-2 (1), E-3 to E-4 (2), E-5 to E-6 (3), E-7 to E-9 (4)
- **Rank category (officer):** O-1 (1), O-2 to O-3 (2), O-4 to O-6 (3)[2]

---

[2]   We did not include general/flag officers (O-7 to O-10) because their numbers were too small for analysis.

- **Time in grade (in years):** 1 year (1), 2 years (2), 3 years (3), and 4 or more years (4).

We ran models separately by service, corps (enlisted, commissioned officer), gender, race/ethnicity, and period. The drawdown models included records for the main drawdown periods: 1990–1998 for the 1990s drawdown; 2003–2008 for the Navy officer drawdown in the mid-2000s; and 2005–2008 for the Air Force drawdown in the mid-2000s. The postdrawdown models used records from the following FYs: 1999–2001 for the 1990s drawdown and 2009–2011 for both the Navy and Air Force mid-2000s drawdowns.

We used discrete-time event history models to net effects of race/ethnicity and gender on separations during drawdown period and after drawdown period.

Each model has the following form:

$$\text{logit}\left(\frac{p_{it}}{1-p_{it}}\right) = a_0 + \sum_{i=2}^{n} a_i YOS_i + \beta_1 x_{it1} + \ldots + \beta_k x_{itk},$$

where p_{it} is the probability of leaving the military in year t; x_{itk} is the workplace characteristic, k, for individual i in year t; $[a_0 + \sum_{i=2}^{n} a_i YOS_i$ is the cumulative continuation rate.

We used a three-step process for our demographic comparisons.

1. Estimate model for each demographic group (e.g., men and women) during drawdown period and postdrawdown period.
2. Estimate counterfactual (adjusted) CCR for minority group (e.g., women, non-Hispanic blacks, or Hispanics, depending on the demographic groups being compared) using the drawdown-period majority group (e.g., men or non-Hispanic whites).
3. Compare observed (unadjusted) CCRs to counterfactual (adjusted) CCRs for demographic groups of interest.

These discrete-time event history models give us group-specific coefficients for the predictors, effectively giving us results as if we estimated a regression model with a series of interactions between demographic indicators (e.g., gender) and other model predictors (e.g., occupational category). We ran separate models for demographic groups for two reasons. First, we find that the regression coefficients are easier to interpret than coefficients with complex interactions. Second, these group-specific regression models allow us to simulate counterfactual scenarios that we describe in the report.

Additional Results

Our modeling approach resulted in 56 models for gender comparisons (4 services × 2 corps × 7 models [2 unadjusted drawdown, 2 unadjusted postdrawdown, 3 adjusted to drawdown male baseline]) and 88 models for racial/ethnic comparisons (4 services × 2 corps × 11 models [3 unadjusted drawdown, 3 unadjusted postdrawdown, 5 adjusted to drawdown white baseline]). To consolidate the results, we created Tables C.11 through C.22 for each service-corps combination (e.g., Army enlisted). These tables offer two types of CCR comparisons:

1. unadjusted CCR comparisons during drawdown and after drawdown (i.e., difference between unadjusted majority and unadjusted female/minority groups)
2. adjusted CCR comparisons to majority drawdown baseline: the difference between unadjusted majority CCRs during drawdown and each of the following three adjusted groups: (1) adjusted female/minority during drawdown, (2) adjusted female/minority postdrawdown, and (3) adjusted majority postdrawdown.

1990s Drawdown

Tables C.11 through C.18 display the results of our modeling approach for the 1990s drawdown in each of the services.

Table C.11
Demographic Differences in CCRs During and After 1990s Drawdown: Army Enlisted

YOS	Unadjusted Comparisons (drawdown/postdrawdown)			Drawdown Male (or White)–Adjusted Group		
	Male-Female	White-Black	White-Hispanic	DD Female/PD Female/PD Male	DD Black/PD Black/PD White	DD Hispanic/PD Hispanic/PD White
1	(0.05/0.07)	(−0.01/−0.03)	(−0.03/−0.05)	(−0.04/−0.04/−0.04)	(−0.05/−0.05/−0.04)	(−0.04/−0.04/−0.04)
2	(0.06/0.10)	(−0.04/−0.04)	(−0.08/−0.09)	(−0.07/−0.06/−0.07)	(−0.08/−0.07/−0.05)	(−0.07/−0.05/−0.05)
3	(0.02/0.05)	(−0.09/−0.06)	(−0.11/−0.10)	(−0.10/−0.08/−0.09)	(−0.11/−0.11/−0.07)	(−0.10/−0.07/−0.07)
4	(−0.01/0.05)	(−0.12/−0.09)	(−0.14/−0.12)	(−0.09/−0.08/−0.09)	(−0.11/−0.10/−0.06)	(−0.09/−0.06/−0.06)
5	(0.00/0.06)	(−0.13/−0.10)	(−0.15/−0.12)	(−0.07/−0.06/−0.07)	(−0.09/−0.09/−0.04)	(−0.08/−0.04/−0.04)
6	(0.00/0.06)	(−0.12/−0.10)	(−0.15/−0.12)	(−0.06/−0.04/−0.05)	(−0.08/−0.07/−0.03)	(−0.06/−0.03/−0.03)
7	(0.00/0.06)	(−0.11/−0.10)	(−0.15/−0.11)	(−0.03/−0.02/−0.03)	(−0.06/−0.05/−0.01)	(−0.05/−0.01/−0.01)
8	(0.01/0.05)	(−0.09/−0.10)	(−0.14/−0.11)	(0.00/0.01/0.00)	(−0.04/−0.03/0.00)	(−0.02/0.00/0.00)
9	(0.01/0.05)	(−0.08/−0.09)	(−0.14/−0.10)	(0.01/0.02/0.02)	(−0.02/−0.02/0.01)	(−0.01/0.01/0.01)
10	(0.01/0.05)	(−0.08/−0.09)	(−0.13/−0.10)	(0.02/0.03/0.03)	(−0.01/−0.01/0.02)	(0.00/0.02/0.02)
11	(0.01/0.04)	(−0.07/−0.09)	(−0.13/−0.09)	(0.03/0.04/0.03)	(−0.01/−0.01/0.02)	(0.00/0.02/0.02)
12	(0.01/0.04)	(−0.07/−0.08)	(−0.12/−0.09)	(0.04/0.04/0.04)	(0.00/0.00/0.02)	(0.01/0.02/0.02)
13	(0.01/0.04)	(−0.06/−0.08)	(−0.11/−0.09)	(0.04/0.05/0.04)	(0.01/0.01/0.03)	(0.01/0.03/0.03)

Table C.11—Continued

YOS	Unadjusted Comparisons (drawdown/postdrawdown)			Drawdown Male (or White)–Adjusted Group		
	Male-Female	White-Black	White-Hispanic	DD Female/PD Female/PD Male	DD Black/PD Black/PD White	DD Hispanic/PD Hispanic/PD White
14	(0.01/0.04)	(−0.05/−0.08)	(−0.11/−0.09)	(0.04/0.05/0.04)	(0.01/0.01/0.03)	(0.01/0.03/0.03)
15	(0.01/0.04)	(−0.05/−0.08)	(−0.10/−0.09)	(0.04/0.05/0.04)	(0.01/0.01/0.03)	(0.02/0.03/0.03)
16	(0.01/0.04)	(−0.04/−0.08)	(−0.10/−0.09)	(0.04/0.05/0.05)	(0.01/0.02/0.03)	(0.02/0.03/0.03)
17	(0.01/0.04)	(−0.04/−0.07)	(−0.09/−0.09)	(0.04/0.05/0.05)	(0.02/0.02/0.03)	(0.02/0.03/0.03)
18	(0.01/0.04)	(−0.04/−0.07)	(−0.09/−0.09)	(0.04/0.05/0.05)	(0.02/0.02/0.03)	(0.02/0.03/0.03)
19	(0.01/0.03)	(−0.04/−0.07)	(−0.09/−0.08)	(0.04/0.05/0.05)	(0.02/0.02/0.03)	(0.02/0.03/0.03)
20	(0.00/0.02)	(−0.03/−0.05)	(−0.05/−0.06)	(0.03/0.03/0.03)	(0.01/0.01/0.02)	(0.01/0.02/0.02)
21	(0.00/0.01)	(−0.02/−0.04)	(−0.04/−0.04)	(0.02/0.02/0.02)	(0.01/0.01/0.01)	(0.01/0.01/0.01)

SOURCE: Analysis of DMDC data on active-duty Army enlisted personnel (FY 1990–2001).

NOTES: Adjusted values reflect female CCRs adjusted to look like male CCRs during the drawdown and minority CCRs adjusted to look like white CCRs during the drawdown. DD stands for "drawdown" period and PD stands for "postdrawdown" period.

Table C.12
Demographic Differences in CCRs During and After 1990s Drawdown: Navy Enlisted

YOS	Unadjusted Comparisons (drawdown/postdrawdown)			Drawdown Male (or White)–Adjusted Group		
	Male-Female	White-Black	White-Hispanic	DD Female / PD Female / PD Male	DD Black / PD Black / PD White	DD Hispanic / PD Hispanic / PD White
1	(0.01/0.00)	(0.00/–0.01)	(–0.01/–0.04)	(–0.05/–0.05/–0.05)	(–0.05/–0.05/–0.05)	(–0.04/–0.04/–0.05)
2	(0.01/–0.01)	(0.01/–0.01)	(–0.01/–0.06)	(–0.09/–.009/–0.10)	(–0.09/–0.09/–0.10)	(–0.08/–0.08/–0.10)
3	(0.01/0.00)	(0.02/0.00)	(–0.01/–0.06)	(–0.09/–0.07/–0.10)	(–0.09/–0.09/–0.11)	(–0.06/–0.06/–0.11)
4	(0.01/–0.02)	(–0.03/–0.04)	(–0.02/–0.07)	(–0.05/–0.03/–0.08)	(–0.05/–0.05/–0.08)	(–0.02/–0.02/–0.08)
5	(0.03/0.02)	(–0.04/–0.07)	(–0.02/–0.07)	(–0.02/.00/–0.05)	(–0.02/–0.02/–0.05)	(0.01/0.01/–0.05)
6	(0.01/0.02)	(–0.08/–0.12)	(–0.04/–0.10)	(0.02/0.03/–0.01)	(0.01/0.01/–0.02)	(0.04/0.04/–0.02)
7	(0.02/0.03)	(–0.08/–0.11)	(–0.04/–0.10)	(0.03/0.05/0.00)	(0.03/0.03/0.00)	(0.05/0.05/0.00)
8	(0.02/0.04)	(–0.07/–0.11)	(–0.04/–0.09)	(0.04/0.06/0.02)	(0.04/0.04/0.01)	(0.06/0.06/0.01)
9	(0.02/0.04)	(–0.07/–0.11)	(–0.04/–0.09)	(0.05/0.06/0.03)	(0.05/0.05/0.02)	(0.07/0.07/0.02)
10	(0.03/0.04)	(–0.06/–0.10)	(–0.04/–0.09)	(0.06/0.07/0.04)	(0.05/0.05/0.03)	(0.07/0.07/0.03)
11	(0.03/0.04)	(–0.06/–0.10)	(–0.03/–0.09)	(0.06/0.07/0.04)	(0.05/0.05/0.03)	(0.07/0.07/0.03)
12	(0.03/0.04)	(–0.06/–0.10)	(–0.03/–0.09)	(0.06/0.07/0.04)	(0.05/0.05/0.04)	(0.07/0.07/0.04)
13	(0.03/0.04)	(–0.05/–0.10)	(–0.03/–0.09)	(0.06/0.07/0.05)	(0.05/0.05/0.04)	(0.07/0.07/0.04)

Table C.12—Continued

YOS	Unadjusted Comparisons (drawdown/postdrawdown)			Drawdown Male (or White)–Adjusted Group		
	Male-Female	White-Black	White-Hispanic	DD Female / PD Female / PD Male	DD Black / PD Black / PD White	DD Hispanic / PD Hispanic / PD White
14	(0.03/0.04)	(−0.05/−0.10)	(−0.03/−0.09)	(0.06/0.07/0.05)	(0.05/0.05/0.04)	(0.07/0.07/0.04)
15	(0.03/0.04)	(−0.05/−0.10)	(−0.03/−0.09)	(0.07/0.07/0.05)	(0.05/0.05/0.04)	(0.06/0.06/0.04)
16	(0.03/0.04)	(−0.05/−0.10)	(−0.03/−0.09)	(0.07/0.07/0.05)	(0.05/0.05/0.04)	(0.06/0.06/0.04)
17	(0.03/0.04)	(−0.05/−0.09)	(−0.03/−0.08)	(0.07/0.07/0.06)	(0.05/0.05/0.04)	(0.06/0.06/0.04)
18	(0.03/0.04)	(−0.05/−.09)	(−0.03/−0.08)	(0.07/0.07/0.06)	(0.05/0.05/0.04)	(0.06/0.06/0.04)
19	(0.03/0.04)	(−0.04/−0.09)	(−0.03/−0.08)	(0.07/0.07/0.06)	(0.05/0.05/0.04)	(0.06/0.06/0.04)
20	(0.02/0.03)	(−0.02/−0.04)	(−0.01/−0.05)	(0.04/0.04/0.03)	(0.03/0.03/0.03)	(0.03/0.03/0.03)
21	(0.01/0.02)	(−0.01/−0.03)	(−0.01/−0.03)	(0.02/0.02/0.02)	(0.02/0.02/0.02)	(0.02/0.02/0.02)

SOURCE: Analysis of DMDC data on active-duty Navy enlisted personnel (FY 1990–2001).

NOTES: Adjusted values reflect female CCRs adjusted to look like male CCRs during the drawdown and minority CCRs adjusted to look like white CCRs during the drawdown. DD stands for "drawdown" period and PD stands for "postdrawdown" period.

Table C.13
Demographic Differences in CCRs During and After 1990s Drawdown: Marine Corps Enlisted

YOS	Unadjusted Comparisons (drawdown/postdrawdown)			Drawdown Male (or White)–Adjusted Group		
	Male-Female	White-Black	White-Hispanic	DD Female/PD Female/PD Male	DD Black/PD Black/PD White	DD Hispanic/PD Hispanic/PD White
1	(0.05/0.04)	(0.00/−0.02)	(−0.02/−0.05)	(−0.04/−0.04/−0.04)	(−0.04/−0.04/−0.04)	(−0.04/−0.04/−0.04)
2	(0.10/0.05)	(0.02/−0.01)	(−0.05/−0.08)	(−0.04/−0.04/−0.04)	(−0.05/−0.05/−0.04)	(−0.03/−0.03/−0.04)
3	(0.12/0.04)	(0.03/−0.01)	(−0.07/−0.09)	(−0.04/−0.03/−0.02)	(−0.04/−0.04/−0.02)	(−0.01/−0.01/−0.02)
4	(0.00/0.02)	(−0.03/−0.06)	(−0.04/−0.07)	(−0.04/−0.04/−0.02)	(−0.04/−0.05/−0.02)	(0.00/−0.01/−0.02)
5	(−0.01/0.00)	(−0.05/−0.10)	(−0.05/−0.09)	(−0.02/−0.01/.00)	(−0.02/−0.03/0.00)	(0.02/0.01/0.00)
6	(−0.02/0.00)	(−0.08/−0.11)	(−0.08/−0.10)	(0.00/0.01/0.02)	(0.00/−0.01/0.02)	(0.03/0.03/0.02)
7	(−0.01/0.01)	(−0.07/−0.10)	(−0.07/−0.09)	(0.01/0.02/0.03)	(0.01/0.01/0.03)	(0.04/0.03/0.03)
8	(−0.01/0.01)	(−0.06/−0.09)	(−0.07/−0.08)	(0.02/0.03/0.04)	(0.02/0.01/0.03)	(0.04/0.04/0.03)
9	(0.00/0.00)	(−0.05/−0.09)	(−0.06/−0.08)	(0.02/0.03/0.04)	(0.02/0.02/0.03)	(0.04/0.04/0.03)
10	(0.00/0.00)	(−0.05/−0.08)	(−0.06/−0.08)	(0.03/0.03/0.04)	(0.02/0.02/0.03)	(0.04/0.04/0.03)
11	(0.00/0.00)	(−0.05/−0.08)	(−0.06/−0.07)	(0.03/0.03/0.04)	(0.02/0.02/0.03)	(0.04/0.04/0.03)
12	(0.00/0.01)	(−0.04/−0.07)	(−0.06/−0.07)	(0.03/0.03/0.04)	(0.02/0.02/0.03)	(0.03/0.03/0.03)
13	(0.00/0.01)	(−0.04/−0.07)	(−0.05/−0.06)	(0.03/0.03/0.04)	(0.02/0.02/0.03)	(0.03/0.03/0.03)

Table C.13—Continued

	Unadjusted Comparisons (drawdown/postdrawdown)			Drawdown Male (or White)–Adjusted Group		
YOS	Male-Female	White-Black	White-Hispanic	DD Female/PD Female/ PD Male	DD Black/PD Black/ PD White	DD Hispanic/PD Hispanic/PD White
14	(0.00/0.01)	(−0.03/−0.06)	(−0.05/−0.06)	(0.03/0.03/0.03)	(0.02/0.02/0.03)	(0.03/0.03/0.03)
15	(0.00/0.01)	(−0.03/−0.06)	(−0.04/−0.06)	(0.03/0.03/0.03)	(0.02/0.02/0.03)	(0.03/0.03/0.03)
16	(0.00/0.01)	(−0.03/−0.06)	(−0.04/−0.06)	(0.03/0.03/0.03)	(0.02/0.02/0.03)	(0.03/0.03/0.03)
17	(0.00/0.01)	(−0.03/−0.06)	(−0.04/−0.06)	(0.03/0.03/0.03)	(0.02/0.02/0.03)	(0.03/0.03/0.03)
18	(0.00/0.01)	(−0.03/−0.06)	(−0.04/−0.06)	(0.03/0.03/0.03)	(0.02/0.02/0.03)	(0.03/0.03/0.03)
19	(0.00/0.01)	(−0.03/−0.06)	(−0.04/−0.06)	(0.03/0.03/0.03)	(0.02/0.02/0.03)	(0.03/0.03/0.03)
20	(0.00/0.01)	(−0.02/−0.04)	(−0.03/−0.04)	(0.02/0.02/0.02)	(0.02/0.02/0.02)	(0.02/0.02/0.02)
21	(0.00/0.01)	(−0.02/−0.03)	(−0.02/−0.03)	(0.02/0.02/0.02)	(001/0.01/0.01)	(0.01/0.01/0.01)

SOURCE: Analysis of DMDC data on active-duty Marine Corps enlisted personnel (FY 1990–2001).

NOTES: Adjusted values reflect female CCRs adjusted to look like male CCRs during the drawdown and minority CCRs adjusted to look like white CCRs during the drawdown. DD stands for "drawdown" period and PD stands for "postdrawdown" period.

Table C.14
Demographic Differences in CCRs During and After 1990s Drawdown: Air Force Enlisted

YOS	Unadjusted Comparisons (drawdown/post-drawdown)			Drawdown Male (or White)-Adjusted Group		
	Male-Female	White-Black	White-Hispanic	DD Female/PD Female/PD Male	DD Black/PD Black/PD White	DD Hispanic/PD Hispanic/PD White
1	(0.02/0.02)	(0.01/0.01)	(−0.02/−0.04)	(−0.04/−0.04/−0.05)	(−0.04/−0.04/−0.04)	(−0.04/−0.04/−0.04)
2	(0.03/0.03)	(0.01/0.02)	(−0.04/−0.06)	(−0.06/−0.06/−0.08)	(−0.07/−0.07/−0.07)	(−0.06/−0.06/−0.07)
3	(0.04/0.04)	(−0.01/0.03)	(−0.07/−0.07)	(−0.07/−0.07/−0.10)	(−0.09/−0.09/−0.10)	(−0.08/−0.07/−0.10)
4	(0.02/0.01)	(−0.08/−0.06)	(−0.10/−0.11)	(−0.10/−0.10/−0.16)	(−0.14/−0.14/−0.15)	(−0.12/−0.10/−0.15)
5	(0.03/0.03)	(−0.09/−0.07)	(−0.10/−0.11)	(−0.09/−0.09/−0.17)	(−0.14/−0.14/−0.15)	(−0.11/−0.09/−0.15)
6	(0.03/0.04)	(−0.09/−0.07)	(−0.11/−0.12)	(−0.09/−0.09/−0.16)	(−0.14/−0.14/−0.15)	(−0.11/−0.09/−0.15)
7	(0.04/0.05)	(−0.09/−0.08)	(−0.11/−0.13)	(−0.08/−0.08/−0.16)	(−0.14/−0.13/−0.15)	(−0.11/−0.08/−0.15)
8	(0.05/0.05)	(−0.09/−0.08)	(−0.11/−0.14)	(−0.06/−0.06/−0.15)	(−0.12/−0.12/−0.14)	(−0.10/−0.07/−0.14)
9	(0.05/0.05)	(−0.08/−0.09)	(−0.11/−0.15)	(−0.04/−0.04/−0.12)	(−0.10/−0.10/−0.11)	(−0.07/−0.05/−0.11)
10	(0.05/0.06)	(−0.07/−0.07)	(−0.10/−0.14)	(0.01/0.01/−0.07)	(−0.06/−0.06/−0.07)	(−0.03/−0.01/−0.07)
11	(0.05/0.06)	(−0.06/−0.07)	(−0.09/−0.14)	(0.04/0.04/−0.04)	(−0.04/−0.04/−0.05)	(−0.01/0.01/−0.05)
12	(0.05/0.06)	(−0.06/−0.07)	(−0.09/−0.14)	(0.05/0.05/−0.02)	(−0.02/−0.02/−0.04)	(0.00/0.02/−0.04)
13	(0.05/0.06)	(−0.06/−0.07)	(−0.09/−0.13)	(0.06/0.06/−0.01)	(−0.01/−0.01/−0.03)	(0.01/0.03/−0.03)

Table C.14—Continued

	Unadjusted Comparisons (drawdown/post-drawdown)			Drawdown Male (or White)–Adjusted Group		
YOS	Male-Female	White-Black	White-Hispanic	DD Female/PD Female/PD Male	DD Black/PD Black/PD White	DD Hispanic/PD Hispanic/PD White
14	(0.05/0.06)	(−0.06/−0.07)	(−0.09/−0.13)	(0.07/0.07/0.00)	(−0.01/−0.01/−0.02)	(0.02/0.03/−0.02)
15	(0.05/0.06)	(−0.06/−0.07)	(−0.08/−0.13)	(0.08/0.08/0.01)	(0.00/0.00/−0.01)	(0.02/0.04/−0.01)
16	(0.05/0.06)	(−0.06/−0.07)	(−0.08/−0.13)	(0.08/0.08/0.02)	(0.01/0.01/0.00)	(0.03/0.05/0.00)
17	(0.05/0.06)	(−0.06/−0.07)	(−0.08/−0.13)	(0.09/0.09/0.03)	(0.02/0.02/0.01)	(0.04/0.05/0.01)
18	(0.05/0.06)	(−0.06/−0.07)	(−0.08/−0.13)	(0.10/0.10/0.05)	(0.03/0.03/0.02)	(0.05/0.06/0.02)
19	(0.05/0.06)	(−0.06/−0.07)	(−0.08/−0.13)	(0.10/0.10/0.06)	(0.04/0.04/0.03)	(0.06/0.07/0.03)
20	(0.03/0.05)	(−0.04/−0.06)	(−0.04/−0.10)	(0.07/0.07/0.06)	(0.04/0.04/0.04)	(0.05/0.05/0.04)
21	(0.02/0.04)	(−0.03/−0.06)	(−0.03/−0.08)	(0.05/0.05/0.05)	(0.04/0.04/0.04)	(0.04/0.04/0.04)

SOURCE: Analysis of DMDC data on active-duty Air Force enlisted personnel (FY 1990–2001).

NOTES: Adjusted values reflect female CCRs adjusted to look like male CCRs during the drawdown and minority CCRs adjusted to look like white CCRs during the drawdown. DD stands for "drawdown" period and PD stands for "postdrawdown" period.

Table C.15
Demographic Differences in CCRs During and After 1990s Drawdown: Army Officer

YOS	Unadjusted Comparisons (drawdown/postdrawdown)			Drawdown Male (or White)–Adjusted Group		
	Male-Female	White-Black	White-Hispanic	DD Female/PD Female/PD Male	DD Black/PD Black/PD White	DD Hispanic/PD Hispanic/PD White
1	(0.01/0.00)	(0.01/0.01)	(0.00/0.00)	(0.01/0.01/0.01)	(0.01/0.01/0.01)	(0.01/0.01/0.01)
2	(0.02/0.01)	(0.01/0.02)	(0.01/-0.01)	(0.06/0.05/0.06)	(0.06/0.06/0.05)	(0.06/0.05/0.05)
3	(0.05/0.02)	(0.00/0.01)	(-0.02/-0.03)	(0.09/0.08/0.09)	(0.10/0.09/0.08)	(0.09/0.07/0.08)
4	(0.11/0.06)	(-0.01/-0.06)	(-0.06/-0.09)	(0.11/0.09/0.10)	(0.12/0.11/0.09)	(0.10/0.08/0.09)
5	(0.12/0.08)	(-0.03/-0.09)	(-0.09/-0.11)	(0.11/0.08/0.10)	(0.12/0.11/0.09)	(0.10/0.08/0.09)
6	(0.13/0.09)	(-0.04/-0.10)	(-0.09/-0.12)	(0.11/0.08/0.10)	(0.12/0.11/0.09)	(0.10/0.07/0.09)
7	(0.15/0.11)	(-0.05/-0.09)	(-0.10/-0.12)	(0.10/0.08/0.10)	(0.12/0.10/0.08)	(0.09/0.07/0.08)
8	(0.15/0.11)	(-0.05/-0.11)	(-0.10/-0.13)	(0.09/0.07/0.09)	(011/0.09/0.07)	(0.08/0.06/0.07)
9	(015/0.11)	(-0.04/-0.11)	(-0.11/-0.12)	(0.07/0.04/0.06)	(0.09/0.07/0.05)	(0.06/0.04/0.05)
10	(0.14/0.11)	(-0.03/-0.12)	(-0.11/-0.12)	(0.05/0.03/0.05)	(0.08/0.06/0.04)	(0.05/0.03/0.04)
11	(0.13/0.11)	(-0.02/-0.11)	(-0.10/-0.12)	(0.03/0.01/0.03)	(0.06/0.04/0.02)	(0.03/0.01/0.02)
12	(0.12/0.11)	(-0.01/-0.11)	(-0.09/-0.12)	(0.02/0.00/0.02)	(0.05/0.03/0.01)	(0.03/0.00/0.01)
13	(0.11/0.11)	(0.00/-0.11)	(-0.09/-0.12)	(0.02/0.00/0.02)	(0.05/0.03/0.01)	(0.02/0.00/0.01)

Table C.15—Continued

YOS	Unadjusted Comparisons (drawdown/postdrawdown)			Drawdown Male (or White)–Adjusted Group		
	Male-Female	White-Black	White-Hispanic	DD Female/PD Female/ PD Male	DD Black/PD Black/PD White	DD Hispanic/PD Hispanic/PD White
14	(0.11/0.11)	(0.00/–0.11)	(–0.09/–0.12)	(0.02/–0.01/0.02)	(0.05/0.03/0.01)	(0.02/0.00/0.01)
15	(0.11/0.11)	(0.00/–0.10)	(–0.08/–0.12)	(0.02/–0.01/0.01)	(0.05/0.03/0.01)	(0.02/0.00/0.01)
16	(0.11/0.11)	(0.00/–0.11)	(–0.08/–0.12)	(0.01/–0.01/0.01)	(0.04/0.03/0.01)	(0.02/–0.01/0.01)
17	(0.11/0.11)	(0.01/–0.10)	(–0.07/–0.12)	(0.01/–0.02/0.01)	(0.04/0.02/0.00)	(0.01/–0.01/0.00)
18	(0.10/0.11)	(0.02/–0.10)	(–0.07/–0.12)	(0.00/–0.02/0.00)	(0.03/0.02/0.00)	(0.01/–0.01/0.00)
19	(0.10/0.11)	(0.02/–0.10)	(–0.07/–0.11)	(0.00/–0.02/0.00)	(0.03/0.02/0.00)	(0.01/–0.01/0.00)
20	(0.08/0.09)	(0.01/–0.07)	(–0.05/–0.09)	(–0.01/–0.03/–0.01)	(0.02/0.01/–0.01)	(0.00/–0.02/–0.01)
21	(0.06/0.07)	(0.01/–0.05)	(–0.04/–0.08)	(–0.01/–0.03/–0.01)	(0.01/0.00/–0.01)	(–0.01/–0.02/–0.01)

SOURCE: Analysis of DMDC data on active-duty Army commissioned officers (FY 1990–2001).

NOTES: Adjusted values reflect female CCRs adjusted to look like male CCRs during the drawdown and minority CCRs adjusted to look like white CCRs during the drawdown. DD stands for "drawdown" period and PD stands for "postdrawdown" period.

Table C.16
Demographic Differences in CCRs During and After 1990s Drawdown: Navy Officer

YOS	Unadjusted Comparisons (drawdown/postdrawdown)			Drawdown Male (or White)–Adjusted Group		
	Male-Female	White-Black	White-Hispanic	DD Female/PD Female/PD Male	DD Black/PD Black/PD White	DD Hispanic/PD Hispanic/PD White
1	(0.00/0.00)	(0.00/0.00)	(0.000/.00)	(0.02/0.02/0.02)	(0.02/0.02/0.02)	(0.02/0.02/0.02)
2	(0.01/0.01)	(0.00/0.00)	(0.00/0.00)	(0.05/0.05/0.05)	(0.05/0.05/0.06)	(0.05/0.04/0.06)
3	(0.04/0.04)	(0.00/0.01)	(0.00/-0.03)	(0.08/0.09/0.10)	(0.08/0.08/0.10)	(0.07/0.06/0.10)
4	(0.06/0.09)	(0.00/0.00)	(-0.03/-0.07)	(0.10/0.11/0.12)	(0.09/0.10/0.12)	(0.08/0.06/0.12)
5	(0.07/0.15)	(-0.03/-0.02)	(-0.05/-0.10)	(0.11/0.12/0.14)	(0.10/0.11/0.14)	(0.09/0.06/0.14)
6	(0.09/0.18)	(-0.04/-0.01)	(-0.05/-0.09)	(0.11/0.13/0.14)	(0.10/0.11/0.14)	(0.09/0.05/0.14)
7	(0.09/0.18)	(-0.06/0.00)	(-0.06/-0.10)	(0.10/0.11/0.12)	(0.08/0.10/013)	(0.07/0.04/0.13)
8	(0.05/0.18)	(-0.08/0.01)	(-0.08/-0.10)	(0.07/0.09/0.10)	(0.05/0.07/0.10)	(0.04/0.01/0.10)
9	(0.03/0.16)	(-0.09/0.00)	(-0.08/-0.12)	(0.03/0.05/0.07)	(0.02/0.04/0.07)	(0.01/-0.03/0.07)
10	(0.02/0.12)	(-0.08/-0.03)	(-0.09/-0.14)	(0.01/0.03/0.05)	(0.00/0.02/0.05)	(-0.01/-0.05/0.05)
11	(0.01/0.10)	(-0.09/-0.03)	(-0.08/-0.15)	(0.00/0.02/0.03)	(-0.01/0.01/0.04)	(-0.02/-0.06/0.04)
12	(0.00/0.09)	(-0.08/-0.03)	(-0.08/-0.15)	(0.00/0.01/0.03)	(-0.01/0.01/0.04)	(-0.02/-0.06/0.04)
13	(0.00/0.09)	(-0.07/-0.02)	(-0.08/-0.13)	(0.00/0.02/0.03)	(-0.01/0.01/0.04)	(-0.02/-0.06/0.04)

Table C.16—Continued

	Unadjusted Comparisons (drawdown/postdrawdown)			Drawdown Male (or White)–Adjusted Group		
YOS	Male-Female	White-Black	White-Hispanic	DD Female/PD Female/PD Male	DD Black/PD Black/PD White	DD Hispanic/PD Hispanic/PD White
14	(0.00/0.08)	(−0.07/−0.02)	(−0.08/−0.13)	(−0.01/0.01/0.03)	(−0.01/0.01/0.04)	(−0.02/−0.06/0.04)
15	(0.01/0.08)	(−0.07/−0.02)	(−0.08/−0.13)	(−0.01/0.01/0.03)	(−0.01/0.00/0.03)	(−0.03/−0.06/0.03)
16	(0.01/0.08)	(−0.07/−0.03)	(−0.08/−0.13)	(−0.01/0.01/0.02)	(−0.02/0.00/0.03)	(−0.03/−0.07/0.03)
17	(0.01/0.08)	(−0.07/−0.02)	(−0.08/−0.13)	(−0.01/0.01/0.02)	(−0.02/0.00/0.03)	(−0.03/−0.07/0.03)
18	(0.01/0.08)	(−0.07/−0.02)	(−0.08/−0.13)	(−0.01/0.00/0.02)	(−0.02/0.00/0.03)	(−0.03/−0.07/0.03)
19	(0.01/0.08)	(−0.06/−0.03)	(−0.08/−0.13)	(−0.02/0.00/0.02)	(−0.02/0.00/0.03)	(−0.03/−0.07/0.03)
20	(0.02/0.08)	(−0.07/−0.05)	(−0.09/−0.11)	(−0.03/−0.02/0.00)	(−0.04/−0.02/0.01)	(−0.05/−0.08/0.01)
21	(0.03/0.07)	(−0.07/−0.05)	(−0.07/−0.10)	(−0.04/−0.03/−0.01)	(−0.05/−0.03/−0.01)	(−0.06/−0.09/−0.01)

SOURCE: Analysis of DMDC data on active-duty Navy commissioned officers (FY 1990–2001).

NOTES: Adjusted values reflect female CCRs adjusted to look like male CCRs during the drawdown and minority CCRs adjusted to look like white CCRs during the drawdown. DD stands for "drawdown" period and PD stands for "postdrawdown" period.

Table C.17
Demographic Differences in CCRs During and After 1990s Drawdown: Marine Corps Officer

YOS	Unadjusted Comparisons (drawdown/postdrawdown)			Drawdown Male (or White)–Adjusted Group		
	Male-Female	White-Black	White-Hispanic	DD Female/PD Female/PD Male	DD Black/PD Black/PD White	DD Hispanic/PD Hispanic/PD White
1	(0.01/0.02)	(0.01/0.03)	(0.00/0.01)	(0.06/0.04/0.05)	(0.05/0.05/0.05)	(0.05/0.04/0.05)
2	(0.02/0.04)	(0.02/0.05)	(0.01/0.01)	(0.09/0.06/0.09)	(0.08/0.08/0.08)	(0.07/0.07/0.08)
3	(0.05/0.06)	(0.04/0.06)	(0.01/0.00)	(0.16/0.10/0.14)	(0.14/0.13/0.14)	(0.12/0.10/0.14)
4	(0.07/0.10)	(0.04/0.04)	(−0.01/0.00)	(0.17/0.08/0.15)	(0.15/0.13/0.15)	(0.12/0.09/0.15)
5	(0.08/0.16)	(0.04/0.02)	(−0.01/−0.03)	(0.17/0.07/0.15)	(0.15/0.13/0.15)	(0.11/0.09/0.15)
6	(0.11/0.18)	(0.04/0.05)	(0.00/0.00)	(0.18/0.07/0.16)	(0.16/0.14/0.16)	(0.12/0.09/0.16)
7	(0.14/0.22)	(0.05/0.06)	(−0.02/0.00)	(0.17/0.06/0.15)	(0.15/0.13/0.14)	(0.11/0.08/0.14)
8	(0.14/0.22)	(0.05/0.06)	(−0.03/−0.01)	(0.15/0.03/0.13)	(0.13/0.10/0.12)	(0.08/0.05/0.12)
9	(0.15/0.22)	(0.05/0.07)	(−0.04/−0.04)	(0.12/0.00/0.10)	(0.10/0.07/0.09)	(0.05/0.02/0.09)
10	(0.14/0.20)	(0.04/0.05)	(−0.04/−0.05)	(0.09/−0.03/0.07)	(0.07/0.05/0.07)	(0.03/−0.01/0.07)
11	(0.12/0.18)	(0.03/0.02)	(−0.06/−0.02)	(0.07/−0.05/0.05)	(0.05/0.02/0.004)	(0.00/−0.03/0.04)
12	(0.11/0.18)	(0.03/0.01)	(−0.07/−0.02)	(0.05/−0.07/0.03)	(0.03/0.00/0.02)	(−0.02/−0.05/0.02)
13	(0.11/0.18)	(0.04/0.02)	(−0.06/−0.01)	(0.04/−0.08/0.01)	(0.02/−0.01/0.01)	(−0.03/−0.06/0.01)

Table C.17—Continued

YOS	Unadjusted Comparisons (drawdown/postdrawdown)			Drawdown Male (or White)–Adjusted Group		
	Male-Female	White-Black	White-Hispanic	DD Female/PD Female/PD Male	DD Black/PD Black/PD White	DD Hispanic/PD Hispanic/PD White
14	(0.10/0.17)	(0.04/0.01)	(–0.07/–0.02)	(0.04/–0.08/0.01)	(0.02/–0.01/001)	(–0.03/–0.06/0.01)
15	(0.11/0.18)	(0.05/0.01)	(–0.06/–0.02)	(0.04/–0.08/0.01)	(0.02/–0.01/0.01)	(–0.03/–0.06/0.01)
16	(0.11/0.17)	(0.05/0.01)	(–0.06/–0.02)	(0.04/–0.08/0.01)	(0.02/–0.01/0.01)	(–0.03/–0.06/0.01)
17	(0.11/0.17)	(0.05/0.01)	(–0.06/–0.01)	(0.03/–0.09/0.01)	(0.01/–0.01/0.01)	(–0.03/–0.06/0.01)
18	(0.10/0.17)	(0.05/0.01)	(–0.06/–0.01)	(0.03/–0.09/0.01)	(0.01/–0.01/0.01)	(–0.03/–0.07/0.01)
19	(0.10/0.16)	(0.05/0.01)	(–0.06/–0.01)	(0.03/–0.09/0.00)	(0.01/–0.02/0.00)	(–0.03/–0.07/0.00)
20	(0.09/0.13)	(0.03/–0.02)	(–0.06/–0.04)	(0.00/–0.11/–0.02)	(–0.01/–0.03/–0.01)	(–0.05/–0.08/–0.01)
21	(0.08/0.12)	(0.01/0.00)	(–0.04/–0.05)	(0.00/–0.11/–0.02)	(–0.01/–0.04/–0.02)	(–0.05/–0.08/–0.02)

SOURCE: Analysis of DMDC data on active-duty Marine Corps commissioned officers (FY 1990–2001).

NOTES: Adjusted values reflect female CCRs adjusted to look like male CCRs during the drawdown and minority CCRs adjusted to look like white CCRs during the drawdown. DD stands for "drawdown" period and PD stands for "postdrawdown" period.

Table C.18
Demographic Differences in CCRs During and After 1990s Drawdown: Air Force Officer

YOS	Unadjusted Comparisons (drawdown/postdrawdown)			Drawdown Male (or White)–Adjusted Group		
	Male-Female	White-Black	White-Hispanic	DD Female/PD Female/ PD Male	DD Black/PD Black/PD White	DD Hispanic/PD Hispanic/PD White
1	(0.01/0.01)	(0.00/0.01)	(−0.01/0.01)	(0.03/0.02/0.02)	(0.03/0.02/0.02)	(0.02/0.02/0.02)
2	(0.02/0.02)	(0.01/0.02)	(−0.01/0.00)	(0.04/0.04/0.04)	(0.05/0.04/0.04)	(0.04/0.03/0.04)
3	(0.08/0.03)	(0.02/0.02)	(−0.04/−0.02)	(0.08/0.08/0.07)	(0.09/0.07/0.07)	(0.08/0.06/0.07)
4	(0.14/0.10)	(0.01/0.05)	(−0.07/−0.11)	(0.12/0.11/0.09)	(0.11/0.09/0.08)	(0.10/0.07/0.08)
5	(0.18/0.13)	(0.02/0.04)	(−0.08/−0.14)	(0.14/0.13/0.11)	(0.14/0.11/0.10)	(0.12/0.08/0.10)
6	(0.20/0.17)	(0.02/0.05)	(−0.09/−0.19)	(0.15/0.14/0.12)	(0.14/0.11/0.10)	(0.13/0.08/0.10)
7	(0.21/0.18)	(0.01/0.04)	(−0.10/−0.21)	(0.15/0.13/0.11)	(0.13/0.10/0.09)	(0.12/0.07/0.09)
8	(0.19/0.19)	(−0.01/0.05)	(−0.10/−0.21)	(0.12/0.11/0.08)	(0.11/0.08/0.07)	(0.09/0.04/0.07)
9	(0.18/0.19)	(−0.01/0.03)	(−0.10/−0.22)	(0.10/0.08/0.05)	(0.09/0.06/0.04)	(0.07/0.02/0.04)
10	(0.18/0.16)	(0.00/0.01)	(−0.10/−0.24)	(0.08/0.06/0.04)	(0.07/0.04/0.03)	(0.05/0.00/0.03)
11	(0.18/0.16)	(0.00/−0.01)	(−0.10/−0.25)	(0.07/0.05/0.02)	(0.06/0.03/0.02)	(0.04/−0.01/0.02)
12	(0.16/0.14)	(0.00/−0.01)	(−0.09/−0.25)	(0.06/0.04/0.01)	(0.06/0.02/0.01)	(0.04/−0.01/0.01)
13	(0.15/0.14)	(0.01/−0.01)	(−0.09/−0.25)	(0.06/0.04/0.01)	(0.06/0.03/0.01)	(0.04/−0.01/0.01)

Table C.18—Continued

	Unadjusted Comparisons (drawdown/postdrawdown)			Drawdown Male (or White)–Adjusted Group		
YOS	Male-Female	White-Black	White-Hispanic	DD Female/PD Female/PD Male	DD Black/PD Black/PD White	DD Hispanic/PD Hispanic/PD White
14	(0.14/0.13)	(0.01/–0.02)	(–0.08/–0.25)	(0.06/0.04/0.01)	(0.06/0.03/0.01)	(0.04/–0.01/0.01)
15	(0.14/0.13)	(0.01/–0.03)	(–0.08/–0.25)	(0.06/0.04/0.01)	(0.05/0.02/0.01)	(0.03/–0.02/0.01)
16	(0.14/0.13)	(0.01/–0.03)	(–0.07/–0.24)	(0.05/0.03/0.00)	(0.05/0.02/0.00)	(0.03/–0.02/0.00)
17	(0.13/0.13)	(0.02/–0.03)	(–0.07/–0.24)	(0.04/0.03/0.00)	(0.04/0.01/0.00)	(0.02/–0.03/0.00)
18	(0.13/0.13)	(0.02/–0.03)	(–0.07/–0.24)	(0.04/0.03/0.00)	(0.04/0.01/0.00)	(0.02/–0.03/0.00)
19	(0.13/0.13)	(0.02/–0.03)	(–0.06/–0.24)	(0.04/0.03/–0.01)	(0.04/0.01/–0.01)	(0.02/–0.03/–0.01)
20	(0.11/0.12)	(0.01/–0.03)	(–0.04/–0.19)	(0.01/0.00/–0.03)	(0.02/–0.01/–0.03)	(0.00/–0.05/–0.03)
21	(0.09/0.10)	(0.01/–0.03)	(–0.04/–0.18)	(0.00/–0.02/–0.04)	(0.00/–0.02/–0.03)	(–0.01/–0.06/–0.03)

SOURCE: Analysis of DMDC data on active-duty Air Force commissioned officers (FY 1990–2001).

NOTES: Adjusted values reflect female CCRs adjusted to look like male CCRs during the drawdown and minority CCRs adjusted to look like white CCRs during the drawdown. DD stands for "drawdown" period and PD stands for "postdrawdown" period.

2000s Drawdown

Table C.19
Demographic Differences in CCRs During 2000s Drawdown: Navy Enlisted

YOS	Unadjusted Comparisons (drawdown only)			Unadjusted Drawdown Male (or White)–Adjusted Group		
	Male-Female	White-Black	White-Hispanic	Drawdown Female	Drawdown Black	Drawdown Hispanic
1	0.02	0.02	−0.05	−0.05	−0.05	−0.05
2	0.02	0.02	−0.07	−0.07	−0.08	−0.07
3	0.03	0.03	−0.08	−0.07	−0.08	−0.06
4	0.03	0.02	−0.05	−0.05	−0.07	−0.04
5	0.04	−0.02	−0.06	−0.01	−0.04	−0.01
6	0.05	−0.05	−0.07	0.02	−0.01	0.03
7	0.06	−0.05	−0.07	0.03	0.01	0.04
8	0.06	−0.02	−0.05	0.05	0.02	0.06
9	0.06	−0.03	−0.05	0.07	0.03	0.07
10	0.06	−0.03	−0.05	0.07	0.04	0.07
11	0.06	−0.03	−0.05	0.08	0.05	0.07
12	0.06	−0.03	−0.05	0.08	0.05	0.07
13	0.06	−0.03	−0.05	0.08	0.05	0.07
14	0.05	−0.02	−0.04	0.09	0.05	0.07
15	0.05	−0.02	−0.04	0.09	0.06	0.07
16	0.05	−0.02	−0.04	0.09	0.06	0.08
17	0.05	−0.02	−0.04	0.09	0.06	0.08
18	0.05	−0.02	−0.04	0.09	0.06	0.08

Table C.19—Continued

| YOS | Unadjusted Comparisons (drawdown only) | | | Unadjusted Drawdown Male (or White)–Adjusted Group | | |
	Male-Female	White-Black	White-Hispanic	Drawdown Female	Drawdown Black	Drawdown Hispanic
19	0.05	−0.02	−0.03	0.09	0.06	0.08
20	0.03	0.00	−0.02	0.05	0.04	0.04
21	0.02	0.00	−0.01	0.04	0.03	0.04

SOURCE: Analysis of DMDC data on active-duty Navy enlisted personnel (FY 2003–2011). Navy enlisted drawdown continued beyond last year of available data, so no postdrawdown analysis was completed.

NOTE: Adjusted values reflect female CCRs adjusted to look like male CCRs during the drawdown and minority CCRs adjusted to look like white CCRs during the drawdown.

Table C.20
Demographic Differences in CCRs During and After 2000s Drawdown: Air Force Enlisted

YOS	Unadjusted Comparisons (drawdown/postdrawdown)			Drawdown Male (or White)-Adjusted Group		
	Male-Female	White-Black	White-Hispanic	DD Female/PD Female/PD Male	DD Black/PD Black/PD White	DD Hispanic/PD Hispanic/PD White
1	(0.04/0.02)	(0.01/0.01)	(−0.04/−0.02)	(−0.05/−0.05/−0.05)	(0.01/0.01/0.01)	(−0.05/−0.05/−0.05)
2	(0.06/0.04)	(0.02/0.03)	(−0.07/−0.03)	(−0.05/−0.06/−0.06)	(0.02/0.02/0.02)	(−0.06/−0.06/−0.05)
3	(0.08/0.06)	(0.03/0.04)	(−0.08/−0.05)	(−0.05/−0.06/−0.06)	(0.03/0.03/0.03)	(−0.06/−0.07/−0.06)
4	(0.10/0.06)	(−0.01/−0.01)	(−0.10/−0.08)	(−0.06/−0.07/−0.07)	(0.05/0.05/0.05)	(−0.07/−0.09/−0.06)
5	(0.11/0.07)	(−0.01/−0.01)	(−0.10/−0.08)	(−0.05/−0.06/−0.06)	(0.07/0.06/0.06)	(−0.06/−0.08/−0.05)
6	(0.06/0.06)	(−0.06/−0.04)	(−0.12/−0.10)	(−0.02/−0.03/−0.03)	(0.07/0.07/0.07)	(−0.04/−0.06/−0.03)
7	(0.06/0.06)	(−0.06/−0.05)	(−0.11/−0.10)	(0.00/−0.02/−0.02)	(0.07/0.07/0.07)	(−0.03/−0.05/−0.01)
8	(0.07/0.08)	(−0.06/−0.05)	(−0.10/−0.09)	(0.02/0.00/0.00)	(0.07/0.06/0.06)	(−0.02/−0.04/0.00)
9	(0.07/0.09)	(−0.06/−0.05)	(−0.10/−0.09)	(0.03/0.01/0.01)	(0.05/0.05/0.05)	(−0.01/−0.04/0.00)
10	(0.07/0.09)	(−0.06/−0.05)	(−0.09/−0.09)	(0.04/0.02/0.02)	(0.04/0.04/0.04)	(−0.01/−0.03/0.01)
11	(0.07/0.09)	(−0.06/−0.05)	(−0.09/−0.09)	(0.04/0.02/0.02)	(0.04/0.04/0.04)	(0.00/−0.03/0.01)
12	(0.07/0.09)	(−0.05/−0.05)	(−0.09/−0.09)	(0.04/0.03/0.03)	(0.04/0.04/0.04)	(0.00/−0.03/0.01)
13	(0.08/0.09)	(−0.05/−0.04)	(−0.09/−0.09)	(0.05/0.03/0.03)	(0.04/0.04/0.04)	(0.00/−0.02/0.02)

Table C.20—Continued

YOS	Unadjusted Comparisons (drawdown/postdrawdown)			Drawdown Male (or White)–Adjusted Group		
	Male-Female	White-Black	White-Hispanic	DD Female/PD Female/PD Male	DD Black/PD Black/PD White	DD Hispanic/PD Hispanic/PD White
14	(0.08/0.09)	(−0.05/−0.04)	(−0.08/−0.09)	(0.05/0.03/0.03)	(0.04/0.03/0.04)	(0.00/−0.02/0.02)
15	(0.08/0.09)	(−0.05/−0.04)	(−0.08/−0.09)	(0.05/0.03/0.03)	(0.04/0.03/0.03)	(0.00/−0.02/0.02)
16	(0.08/0.09)	(−0.05/−0.04)	(−0.08/−0.09)	(0.05/0.03/0.04)	(0.04/0.03/0.03)	(0.00/−0.02/0.02)
17	(0.08/0.09)	(−0.05/−0.04)	(−0.08/−0.09)	(0.05/0.04/0.04)	(0.04/0.03/0.03)	(0.01/−0.02/0.02)
18	(0.08/0.09)	(−0.05/−0.04)	(−0.08/−0.09)	(0.05/0.04/0.04)	(0.04/0.03/0.03)	(0.01/−0.02/0.02)
19	(0.08/0.09)	(−0.05/−0.04)	(−0.08/−0.09)	(0.06/0.04/0.04)	(0.04/0.03/0.03)	(0.01/−0.02/0.02)
20	(0.06/0.08)	(−0.05/−0.04)	(−0.07/−0.07)	(0.07/0.05/0.05)	(0.02/0.02/0.02)	(0.03/0.01/0.04)
21	(0.05/0.07)	(−0.05/−0.04)	(−0.06/−0.07)	(0.07/0.06/0.06)	(0.02/0.01/0.01)	(0.03/0.02/0.04)

SOURCE: Analysis of DMDC data on active-duty Air Force enlisted personnel (FY 2005–2011). DD stands for "drawdown" period and PD stands for "postdrawdown" period.

NOTE: Adjusted values reflect female CCRs adjusted to look like male CCRs during the drawdown and minority CCRs adjusted to look like white CCRs during the drawdown.

Table C.21
Demographic Differences in CCRs During and After 2000s Drawdown: Navy Officer

YOS	Unadjusted Comparisons (drawdown/postdrawdown)			Drawdown Male (or White)–Adjusted Group		
	Male-Female	White-Black	White-Hispanic	DD Female / PD Female / PD Male	DD Black/PD Black/PD White	DD Hispanic/PD Hispanic/PD White
1	(0.00/−0.01)	(0.00/−0.02)	(0.00/−0.01)	(0.01/0.01/0.01)	(0.01/0.01/0.01)	(0.01/0.01/0.01)
2	(0.01/−0.02)	(0.01/−0.03)	(0.01/0.00)	(0.02/0.02/0.02)	(0.02/0.02/0.02)	(0.02/0.02/0.02)
3	(0.03/0.00)	(0.03/0.00)	(0.01/0.01)	(0.04/0.04/0.04)	(0.03/0.03/0.03)	(0.03/0.03/0.03)
4	(0.10/0.07)	(0.03/0.01)	(0.00/0.00)	(0.06/0.05/0.06)	(0.05/0.05/0.05)	(0.04/0.04/0.05)
5	(0.15/0.11)	(0.02/−0.01)	(−0.01/0.00)	(0.07/0.07/0.07)	(0.07/0.06/0.06)	(0.05/0.06/0.06)
6	(0.17/0.13)	(0.03/−0.01)	(−0.01/−0.01)	(0.08/0.08/0.08)	(0.07/0.07/0.07)	(0.06/0.06/0.07)
7	(0.18/0.13)	(0.03/0.00)	(0.00/−0.02)	(0.08/0.08/0.08)	(0.07/0.07/0.07)	(0.06/0.06/0.07)
8	(0.18/0.14)	(0.04/−0.01)	(0.01/−0.03)	(0.08/0.07/0.08)	(0.07/0.06/0.06)	(0.05/0.05/0.06)
9	(0.18/0.14)	(0.02/−0.03)	(0.01/−0.05)	(0.06/0.05/0.06)	(0.05/0.05/0.05)	(0.04/0.04/0.05)
10	(0.17/0.15)	(0.02/−0.04)	(0.00/−0.05)	(0.05/0.04/0.05)	(0.04/0.04/0.04)	(0.03/0.03/0.04)
11	(0.17/0.14)	(0.02/−0.04)	(0.00/−0.05)	(0.05/0.04/0.05)	(0.04/0.04/0.04)	(0.03/0.03/0.04)
12	(0.16/0.13)	(0.02/−0.04)	(0.00/−0.05)	(0.05/0.04/0.05)	(0.04/0.04/0.04)	(0.03/0.03/0.04)
13	(0.16/0.13)	(0.01/−0.04)	(0.00/−0.05)	(0.05/0.04/0.05)	(0.04/0.04/0.04)	(0.03/0.03/0.04)
14	(0.16/0.13)	(0.01/−0.03)	(0.00/−0.05)	(0.05/0.04/0.05)	(0.04/0.03/0.04)	(0.02/0.03/0.04)

Table C.21—Continued

YOS	Unadjusted Comparisons (drawdown/postdrawdown)			Drawdown Male (or White)–Adjusted Group		
	Male-Female	White-Black	White-Hispanic	DD Female/PD Female/PD Male	DD Black/PD Black/PD White	DD Hispanic/PD Hispanic/PD White
15	(0.16/0.13)	(0.02/–0.03)	(0.01/–0.05)	(0.04/0.04/0.04)	(0.04/0.03/0.03)	(0.02/0.02/0.03)
16	(0.16/0.13)	(0.02/–0.03)	(0.01/–0.05)	(0.04/0.04/0.04)	(0.04/0.03/0.03)	(0.02/0.02/0.03)
17	(0.16/0.13)	(0.02/–0.03)	(0.01/–0.05)	(0.04/0.04/0.04)	(0.04/0.03/0.03)	(0.02/0.02/0.03)
18	(0.16/0.13)	(0.02/–0.02)	(0.00/–0.04)	(0.04/0.04/0.04)	(0.04/0.03/0.03)	(0.02/0.02/0.03)
19	(0.16/0.13)	(0.02/–0.02)	(0.00/–0.04)	(0.04/0.04/0.04)	(0.04/0.03/0.03)	(0.02/0.02/0.03)
20	(0.15/0.13)	(0.00/–0.03)	(–0.01/–0.04)	(0.03/0.02/0.03)	(0.02/0.02/0.02)	(0.01/0.01/0.02)
21	(0.14/0.13)	(–0.01/–0.04)	(–0.01/–0.04)	(0.02/0.01/0.02)	(0.02/0.01/0.01)	(0.00/0.00/0.01)

SOURCE: Analysis of DMDC data on active-duty Navy commissioned officers (FY 2003–2011). DD stands for "drawdown" period and PD stands for "postdrawdown" period.

NOTE: Adjusted values reflect female CCRs adjusted to look like male CCRs during the drawdown and minority CCRs adjusted to look like white CCRs during the drawdown.

Table C.22
Demographic Differences in CCRs During and After 2000s Drawdown: Air Force Officer

YOS	Unadjusted Comparisons (drawdown/postdrawdown)			Drawdown Male (or White)–Adjusted Group		
	Male-Female	White-Black	White-Hispanic	DD Female/PD Female/PD Male	DD Black/PD Black/PD White	DD Hispanic/PD Hispanic/PD White
1	(0.01/0.00)	(0.00/0.00)	(−0.01/−0.01)	(0.02/0.02/0.02)	(0.02/0.02/0.01)	(0.01/0.02/0.01)
2	(0.03/0.00)	(0.02/0.00)	(−0.01/0.00)	(0.05/0.05/0.04)	(0.05/0.05/0.04)	(0.04/0.05/0.04)
3	(0.08/0.03)	(0.04/0.00)	(−0.01/0.00)	(0.08/0.08/0.07)	(0.08/0.08/0.06)	(0.07/0.08/0.06)
4	(0.13/0.09)	(0.08/0.00)	(−0.01/0.00)	(0.11/0.11/0.10)	(0.11/0.11/0.08)	(0.09/0.11/0.08)
5	(0.16/0.14)	(0.09/0.00)	(0.00/−0.02)	(0.13/0.13/0.12)	(0.14/0.14/0.10)	(0.11/0.14/0.10)
6	(0.21/0.17)	(0.11/0.00)	(0.01/−0.02)	(0.15/0.15/0.14)	(0.15/0.16/0.11)	(0.13/0.15/0.11)
7	(0.23/0.21)	(0.12/0.03)	(0.02/−0.01)	(0.16/0.15/0.14)	(0.16/0.16/0.12)	(0.13/0.16/0.12)
8	(0.24/0.24)	(0.14/0.07)	(0.03/0.00)	(0.15/0.15/0.14)	(0.16/0.16/0.11)	(0.13/0.16/0.11)
9	(0.24/0.25)	(0.12/0.07)	(0.02/0.01)	(0.14/0.14/0.13)	(0.14/0.15/0.10)	(0.12/0.14/0.10)
10	(0.22/0.26)	(0.11/0.08)	(0.02/0.03)	(0.13/0.12/0.11)	(0.13/0.14/0.09)	(0.10/0.13/0.09)
11	(0.21/0.26)	(0.11/0.07)	(0.02/0.03)	(0.12/0.11/0.10)	(0.12/0.13/0.08)	(0.09/0.12/0.08)
12	(0.21/0.25)	(0.10/0.06)	(0.03/0.02)	(0.11/0.11/0.09)	(0.11/0.12/0.07)	(0.09/0.11/0.07)
13	(0.20/0.25)	(0.10/0.06)	(0.03/0.02)	(0.10/0.10/0.09)	(0.11/0.11/0.07)	(0.08/0.11/0.07)

Table C.22—Continued

	Unadjusted Comparisons (drawdown/ postdrawdown)			Drawdown Male (or White)–Adjusted Group		
YOS	Male-Female	White-Black	White-Hispanic	DD Female/PD Female/ PD Male	DD Black/PD Black/PD White	DD Hispanic/PD Hispanic/PD White
14	(0.20/0.25)	(0.10/0.06)	(0.03/0.02)	(0.10/0.10/0.08)	(0.11/0.11/0.07)	(0.08/0.11/0.07)
15	(0.20/0.24)	(0.10/0.07)	(0.03/0.02)	(0.10/0.09/0.08)	(0.10/0.11/0.06)	(0.08/0.10/0.06)
16	(0.20/0.23)	(0.09/0.07)	(0.03/0.03)	(0.10/0.09/0.08)	(0.10/0.11/0.06)	(0.08/0.10/0.06)
17	(0.20/0.23)	(0.10/0.07)	(0.03/0.03)	(0.09/0.09/0.08)	(0.10/0.11/0.06)	(0.07/0.10/0.06)
18	(0.20/0.23)	(0.10/0.07)	(0.03/0.03)	(0.09/0.09/0.08)	(0.10/0.11/0.06)	(0.07/0.10/0.06)
19	(0.20/0.23)	(0.10/0.08)	(0.03/0.03)	(0.09/0.09/0.08)	(0.10/0.11/0.06)	(0.07/0.10/0.06)
20	(0.15/0.20)	(0.06/0.07)	(0.02/0.05)	(0.04/0.04/0.03)	(0.05/0.06/0.01)	(0.03/0.05/0.01)
21	(0.13/0.17)	(0.04/0.04)	(0.01/0.04)	(0.02/0.02/0.00)	(0.03/0.03/0.00)	(0.01/0.03/0.00)

SOURCE: Analysis of DMDC data on active-duty Air Force commissioned officers (FY 2005–2011). DD stands for "drawdown" period and PD stands for "postdrawdown" period.

NOTE: Adjusted values reflect female CCRs adjusted to look like male CCRs during the drawdown and minority CCRs adjusted to look like white CCRs during the drawdown.

APPENDIX D

Overview of Tools Available for Recent Drawdown

This appendix provides more details on our review of tools available for recent drawdowns. Our review began by identifying legal authorities and DoD policies that govern the use of drawdown tools across DoD. Although we focus primarily on tools authorized in law or OSD policy, we provide service-specific examples of certain tools. Our review is not exhaustive given the limitations of publicly available data and that the use of drawdown tools continues to evolve. For example, authorities can go out of date and the services modify their drawdown plans, including what and how they use certain tools. However, we offer key themes that encompass the methods by which drawdown tools identify service members for separation.

Legal Authorities and DoD Policies

Title 10 of U.S. Code, entitled Armed Forces, provides the legal authority for the majority of the service's drawdown tools, outlining the legal basis for governing the Armed Forces. The NDAA includes legal authority for certain drawdown tools as well. Additionally, DoD or service-level issuances outline policies for many of the currently available drawdown tools. These include DoD instructions and service policies that outline separation policies for service members at the DoD level or tailored for the specific service. Table D.1 provides examples of authorities available at different levels of law and policy.

Table D.1
Sample Legal Authorities and DoD/Service Policies

Legal Authority	Title/Subject	Sample Drawdown Tool Authorized
Title 10, Section 638	Selective Early Retirement	• Early Discharge Authority • Voluntary Retirement Incentive • Selective Early Retirement Board
Title 10, Section 1175	Voluntary Separation Incentive	• Voluntary Separation Pay

DoD Policy	Title/Subject	Sample Drawdown Tool Authorized
DoD Instruction 1332.30	Separation of Regular and Reserve Commissioned Officers	• Force Shaping Board
DoD Instruction 1332.32	Selective Early Retirement of Officers on an Active Duty List and the Reserve Active Status List and Selective Early Removal of Officers from the Reserve Active Status List	• Selective Early Retirement Board

Service Policy	Title/Subject	Sample Drawdown Tool Authorized
OPNAV Instruction 1811.3A	Voluntary Retirement and Transfer to the Fleet Reserve of Members of the Navy Serving on Active Duty	• Time in Grade Waivers
Army Directive 2013-14	Temporary Early Retirement Authority	• Temporary Early Retirement Authority
Air Force Instruction 36-2107	Active Duty Service Commitments	• Active Duty Service Commitment Waivers

Overview of Drawdown Tools

As of FY 2014, a number of drawdown tools were available to the services for force shaping. These tools vary in their approach to separating service members. For instance, some drawdown tools target either the officer or enlisted ranks, while others can be used across the force. Additionally, drawdown tools can be voluntary by inducing service members to leave military service early or involuntary by forcing service members to leave. As described earlier in the chapter, a board process that selects individuals for separation is an involuntary separation tool. Voluntary separation tools, like VSP, often require significant resources because they usually involve monetary separation incentives. Thus, service budget limitations could encourage the use of involuntary drawdown tools.

In our review of available drawdown tools, certain key themes emerged regarding the way in which tools target populations for separation. We found that current drawdown tools identify service members for separation from the military based on three major categories: experience, occupation, and quality. Experience-based drawdown tools focus on service members' separations according to years of service and/or rank. The majority of drawdown tools currently available are experience-based, with many offering incentives or opportunities for early retirement. Occupation-based drawdown tools identify members for separation based on occupation or career field, primarily focusing on occupations currently overmanned. Quality-driven drawdown tools seek to retain only the highest-quality service members, identifying those service members for separation who are underperformers, have disciplinary issues, and/or are not promotable (e.g., been already passed over for promotion).

There are certain tools or circumstances in which more than one key theme may be relevant when the tool is used. For instance, a tool may target a population for separation based on rank or years of service, but then select individuals for separation within that targeted pool based on quality. Because of the indistinct uses of some tools, we will not categorize each available tool by key theme. Rather, we will treat these key themes as the major levers that the services can use in force management and discuss them in greater detail by using examples of drawdown tools.

Examples of Drawdown Tools

This section outlines specific drawdown tools to serve as examples of how the key themes or levers may be currently used.

Legal Authority or OSD Policy–Driven Tools
As discussed, Title 10 is the primary legal authority for DoD draw-down tools. The following examples are outlined in various sections of Title 10.

- **Selective Early Retirement Board (SERB) and Enhanced SERB (E-SERB).** SERBs identify officers for involuntary early retirement based on year group and competitive category. According-ing to Title 10, Section 638, SERBs select separations from a pool of officers meeting the following requirements:
  – O-5: has been nonselected for promotion twice or more and is not currently recommended for promotion
  – O-6: has served four or more years in current grade and is not currently recommended for promotion
  – O-7: has served 3.5 or more years in current grade and is not currently recommended for promotion
  – O-8: has served 3.5 or more years in current grade.

Officers in the rank of O-5 and O-6 can be considered by a SERB only once every five years in the same grade. Title 10 also limits the number of early retirement separations through SERBs to a maximum of 30 percent of officers in each grade and competitive category. The E-SERB expands the scope of the SERB and provides greater flexibil-ity for its use as a force management tool. Specifically, the E-SERB expands the pool of officers who can be considered for discharge, including:
  – O-5 officers with only a one-time nonselection for promotion who are not currently selected for promotion
  – O-6 officers who have served two or more (rather than at least four) years in current grade and are not currently recommended for promotion
  – officers below the rank of O-5 who have served at least one year in their current grade, are not currently selected for promotion, and are not eligible for retirement within the next two years.

Additionally, the E-SERB suspends the five-year limitation for board consideration, meaning officers who meet requirements may face a board every year. Finally, the E-SERB also allows the board to target occupational specialties for discharge.

The SERB and E-SERB are example of tools that exhibit all three key themes in the manner in which they target officers for separation. The tools have experienced-based elements in terms of targeting certain ranks and required years of service. Quality themes emerge as nonselection for promotion becomes a factor for consideration. Additionally, occupational specialties are a basis for identifying officers for separation in the E-SERB. Clearly, while each of these three themes is present in the identification in the pool for consideration for a SERB or E-SERB, the services have flexibility in which key theme will primarily drive the final discharge decisions from that pool of selected officers.

- **TERA, VSP, and Voluntary Retirement Incentive (VRI).** TERA, VSP, and VRI are examples of voluntary drawdown tools authorized by Title 10. TERA and VSP are available for both the officer and enlisted ranks, while VRI is used on only the officer population. TERA authorizes early retirement pay for service members with more than 15 years of service, but not yet 20 years of service. The amount of retirement pay received is based on service members' years of service,[1] with the service funding retirement pay from date of retirement until 20 years of service is reached, at which point payment transitions to the Military Retirement Fund (MRF). VSP extends the authority to offer separation pay to service members with more than six years of service, but not yet 20 years of service.[2] VRI is a program incentivizing retirement-eligible officers with 20 to 29 years of service to retire by offering a lump sum payment at retirement.[3]

---

[1] Annual basic pay × YOS × 2.5% × reduction factor.

[2] Amount not to exceed four times full separation pay (10% of [12 × monthly base pay × YOS]).

[3] This payment cannot exceed 12 times the amount of an officer's monthly basic pay at the time of retirement and is limited to 675 officers across DoD.

While these three tools are driven by the key theme of experience as service members are identified for separation eligibility based on years of service, other themes may be involved in how these tools are implemented. For example, service members only in certain career fields that a service is aiming to reduce may be offered the opportunity to apply for TERA, VSP, or VRI. Also, officers being considered by a SERB, for instance, may be given the opportunity to apply for TERA. Not only do multiple key themes come into play for certain drawdown tools, but more than one drawdown tool is often used in conjunction with others in force management.

Service-Specific Tools

Within legal guidelines and OSD policy regulations, services often establish specific drawdown tools to meet their particular force management needs. These tools are typically based on a broader drawdown tool and then tailored for specific service requirements. Often services may differ in their implementation of these tools, perhaps differing in which key element drives selections or whether a tool is used for voluntary or involuntary separations.

- **Date of Separation (DOS) Rollback.** The DOS Rollback program is an Air Force program, which is based on the broader Early Discharge Authority.[4] Both tools are intended for use with the enlisted ranks. Similar to the Early Discharge Authority, the DOS Rollback program identifies enlisted Airmen for discharge up to 12 months prior to the expiration of their enlistment term. Specifically, enlisted Airmen up to the rank of E-8 can be involuntarily separated under this program if they have refused permanent change of station, temporary duty, training, retraining, or professional military education (Losey, 2013a). DOS Rollback has typically identified Airmen for early discharge based on negative quality indicators. The program may also select Airmen for early discharge based on occupation as well. Other examples of

---

[4] The Early Discharge Authority extends the period prior to the expiration of an enlistment term from three months to one year, during which a service member may be discharged without loss of benefits.

service-specific tailoring of the Early Discharge Authority include using this authority as a voluntary rather than involuntary tool in some services.

- **Qualitative Service Program (QSP).** QSP is an Army program similar to the Selective Re-enlistment Opportunity program[5] and Enlisted Retention Boards (ERB)[6] used by other services. These programs are intended for use on enlisted service members and involve involuntary separations. QSP focuses on retention screenings of noncommissioned officers (NCOs) through QSP panels (Wiggins, 2012). QSP includes three separate selection programs:
  - Qualitative Management Program: reviews senior NCOs for denial of continued service based on performance
  - Over-Strength Qualitative Service Program: identifies NCOs in an overmanned military occupational specialty (MOS) for separation
  - Promotion Stagnation Qualitative Service Program: targets enlisted members in specific MOSs that have not been promoted to a certain rank and have produced a stagnant promotion situation within those MOSs.

As evidenced through these three subprograms, QSP relies on multiple key themes to drive its separation decisions, with quality as the overarching foundation of the program.

Summary

Through the examples outlined above, it is clear that the services have a multitude of tools to aid in force management. These tools vary in terms of their intended targets (e.g., officer or enlisted ranks), whether separations are voluntary or involuntary (i.e., separation decisions are

---

[5]   Allows the services to deny reenlistment to enlisted service members in overmanned occupations or who have performance issues.

[6]   Allows for the selection of enlisted service members for separation through denial of reenlistment in overmanned occupations and grades.

driven by service or service member), and the key themes that guide the basis for the separation decision (i.e., separation decisions based on experience, occupation, or quality). When the services implement drawdown tools, they tend to use them in combinations that simultaneously affect more than one theme driving separation decisions. Furthermore, services regularly tailor force management authorities to create tools to meet their specific needs.

Despite the varying methods of implementing drawdown tools, overall trends do emerge. For example, across the services, experienced-based voluntary separation tools most often target officers nearing retirement. Involuntary drawdown tools generally have a quality component and can target a broader range of service members in terms of rank and years of service. The overarching legal authorities provide the framework for how the services can use force management tools, thus allowing these overall trends to emerge, given the variations in drawdown tool implementation.

References

Agresti, A., *An Introduction to Categorical Data Analysis*, New York: Wiley, 1996.

Alexander, D., and A. Shalal, "Budget Cuts to Slash U.S. Army to Smallest Since Before World War Two," Reuters, February 24, 2014. As of August 12, 2015:
http://www.reuters.com/article/2014/02/24/
us-usa-defense-budget-idUSBREA1N1IO20140224

Alvin v. United States, 50 Fed. Cl. 295 (2001).

Army Directive 2013-14, *Temporary Early Retirement Authority*, Washington, D.C., Secretary of the Army, June 10, 2013.

Army Times staff writers, "In 2015, Army Will Lose Nearly 20,000 Soldiers in Drawdown," *Army Times*, December 27, 2014. As of August 12, 2015:
http://www.armytimes.com/story/military/careers/
army/2014/12/26/2015-drawdown-year-ahead/20860491/

Asch, B. J., and J. T. Warner, *An Examination of the Effects of Voluntary Separation Incentives*, Santa Monica, Calif.: RAND Corporation, MR-859-OSD, 2001.

Baker v. United States, 34 Fed. Cl. 645 (1995).

Berkley v. United States, 48 Fed. Cl. 361 (2002).

———, 287 F.3d 1076 (Fed. Cir. 2002).

———, 59 Fed. Cl. 675 (2004).

Bowling, T. P., *Federal Downsizing: The Status of Agencies' Workforce Reduction Efforts*, Testimony Before the Subcommittee on Civil Service, Committee on Government Reform and Oversight, House of Representatives, Washington, D.C.: U.S. General Accounting Office, GAO/T-GGD-96-124, May 23, 1996. As of August 12, 2015:
http://www.gao.gov/assets/110/106507.pdf

Brinkerhoff, J. R., "Army TOE and TDA Personnel: FY 1979–FY 1999," Alexandria, Va.: Institute for Defense Analyses, IDA Document D-2460, 2000.

Brostek, M., and B. Holman, *Human Capital: Strategic Approach Should Guide DOD Civilian Workforce Management*, Washington, D.C.: U.S. General Accounting Office, GAO/T-GGD/NSIAD-00-120, March 9, 2000.

Brown, E. A., and M. Millham, "Army Identifies Hundreds for Separation," *Stars and Stripes*, February 12, 2014. As of August 12, 2015: http://www.stripes.com/news/army-identifies-hundreds-for-separation-1.267173

CBO—*See* Congressional Budget Office

Chapman, S., "Civilian Drawdown, Hard and Fast," *Air Force Magazine*, January 1996, p. 28.

Christian v. United States, 46 Fed. Cl. 793 (2000).

Congressional Budget Office, *The Drawdown of the Military Officer Corps*, Washington, D.C.: Congressional Budget Office, November 1999. As of August 20, 2015: https://www.cbo.gov/sites/default/files/106th-congress-1999-2000/reports/drawdown.pdf

———, *Recruiting, Retention, and Future Levels of Military Personnel*, Washington, D.C.: Congressional Budget Office, 2006. As of August 12, 2015: http://www.cbo.gov/sites/default/files/10-05-recruiting.pdf

Congressional Commission on Military Training and Gender-Related Issues, *Final Report: Findings and Recommendations*, Volume I, July 1999. As of August 12, 2015: http://www.dtic.mil/dtfs/doc_research/p18_16v1.pdf

Conley, R. E., A. A. Robbert, J. G. Bolten, M. Carrillo, and H. G. Massey, *Maintaining the Balance Between Manpower, Skill Levels, and PERSTEMPO*, Santa Monica, Calif.: RAND Corporation, MG-492-AF, 2006. As of August 12, 2015: http://www.rand.org/pubs/monographs/MG492.html

Daniel, L., "Active-Duty Downsizing Should Benefit Reserve Forces, Board Says," *American Forces Press Service*, March 12, 2012. As of September 22, 2015: http://www.nationalguard.mil/News/ArticleView/tabid/5563/Article/575968/active-duty-downsizing-should-benefit-reserve-forces.aspx

DoD—*See* U.S. Department of Defense

"DoD Makes It Official: Budget Cuts Will Shrink Army to 420,000 Soldiers," *InsideDefense.com*, Vol. 30, No. 3, January 10, 2014. As of August 12, 2015: http://insidedefense.com/inside-pentagon/dod-makes-it-official-budget-cuts-will-shrink-army-420000-soldiers

DoDD—*See* U.S. Department of Defense, Department of Defense Directive

EEOC—*See* Equal Employment Opportunity Commission.

Equal Employment Opportunity Commission, *Uniform Guidelines on Employee Selection Procedures (1978)*, Section 60-3, 43 FR 38295, August 25, 1978.

Faram, M. D., "Navy: Enlisted Retention Boards Cut 2,947," *Navy Times*, December 2, 2011. As of August 12, 2015:
http://www.navytimes.com/article/20111202/NEWS/112020323/
Navy-Enlisted-retention-boards-cut-2-947

Feickert, A., *Army Drawdown and Restructuring: Background and Issues for Congress*, Washington, D.C.: Congressional Research Service, R42493, 2014a. As of August 12, 2015:
http://fpc.state.gov/documents/organization/223457.pdf

————, *Marine Corps Drawdown, Force Structure Initiatives, and Roles and Missions: Background and Issues for Congress*, Washington, D.C.: Congressional Research Service, R43355, 2014b. As of August 12, 2015:
http://fpc.state.gov/documents/organization/221267.pdf

Fisher v. University of Texas, 570 U.S. __ (2013), Docket No. 11-345.

GAO—*See* U.S. General Accounting Office or U.S. Government Accountability Office.

Gibson, H. O., *The Total Army Competitive Category Optimization Model: Analysis of U.S. Army Officer Accessions and Promotions,* thesis, Monterey, Calif.: Naval Postgraduate School, 2007.

Gildea, D., "AF to Convene CMSgt Retention Board in June," U.S. Air Force website, December 11, 2013. As of August 12, 2015:
http://www.af.mil/News/ArticleDisplay/tabid/223/Article/467712/af-to-convene-cmsgt-retention-board-in-june.aspx

————, "Enlisted Quality Force Review Board to Be Held in May," U.S. Air Force website, January 3, 2014. As of August 12, 2015:
http://www.af.mil/News/ArticleDisplay/tabid/223/Article/467855/enlisted-quality-force-review-board-to-be-held-in-may.aspx

Gillert, D. J., "Voluntary RIF Tops New Civilian Drawdown Options," *American Forces Press Service*, May 3, 1996. As of September 22, 2015:
http://archive.defense.gov/news/newsarticle.aspx?id=40836

Gore, L., "Pentagon Aims to Shed 50,000 Civilian Jobs in the Next Five Years," *The Huntsville Times*, April 11, 2013. As of August 12, 2015:
http://blog.al.com/breaking/2013/04/pentagon_aims_to_shed_50000_ci.html

Grutter v. Bollinger, 539 U.S. 306 (2003).

Hansen, M. L., and S. Nataraj, *Expectations About Civilian Labor Markets and Army Officer Retention*, Santa Monica, Calif.: RAND Corporation, MG-1123-A, 2011.

Hansen, M. L., and J. W. Wenger, *Why Do Pay Elasticities Differ?* Alexandria, Va.: CNA, CRM D0005644.A2/Final, 2002.

Henning, C. A., *Military Retirement: Major Legislative Issues*, Congressional Research Service, Order Code IB85159, 2006. As of August 12, 2015: https://www.fas.org/sgp/crs/natsec/IB85159.pdf

Holliman, S. D., "Civilian Personnel: Employment Levels, Separations, Transition Programs and Downsizing Strategy," Annex K to *Adjusting to the Drawdown*, Report of the Defense Conversion Commission, Washington, D.C.: Department of Defense, February 1993.

Hosek, S. D., P. Tiemeyer, M. R. Kilburn, D. A. Strong, S. Ducksworth, and R. Ray, *Minority and Gender Differences in Officer Career Progression*, Santa Monica, Calif.: RAND Corporation, MR-1184-OSD, 2001. As of August 12, 2015: http://www.rand.org/pubs/monograph_reports/MR1184.html

Jansen, D. J., D. F. Burrelli, L. Kapp, and C. A. Theohary, *FY2014 National Defense Authorization Act: Selected Military Personnel Issues*, Congressional Research Service, R43184, February 24, 2014. As of August 12, 2015: https://www.hsdl.org/?view&did=750604

Jordan, B., "Guard Association Hits Army Chief over Remarks," Military.com, January 14, 2014. As of August 12, 2015: http://www.military.com/daily-news/2014/01/14/guard-association-hits-army-chief-over-remarks.html

Kennedy, H., "At War, Navy Finds New Uses for Reserve Forces," *National Defense*, September 2004. As of August 12, 2015: http://www.nationaldefensemagazine.org/ARCHIVE/2004/SEPTEMBER/Pages/At_War3412.aspx

Kingsbury, N., *Federal Personnel: The EEO Implications of Reductions-in-Force*, Testimony Before the Subcommittee on Civil Service and Subcommittee and Compensation and Employee Benefits, Committee on Post Office and Civil Service, House of Representatives, Washington, D.C.: U.S. General Accounting Office, GAO/T-GGD-94-87, February 1, 1994.

Kirby, M. A., *A Multivariate Analysis of the Effects of the VSI/SSB Separation Program on Navy Enlisted Personnel*, thesis, Monterey, Calif.: Naval Postgraduate School, 1993.

Lamothe, D., "Re-enlistment Changes Coming this Summer," *Marine Corps Times*, May 31, 2011. As of August 12, 2015: http://archive.marinecorpstimes.com/article/20110531/NEWS/105310339/Re-enlistment-changes-coming-summer

Lewis, G. B., "The Impact of Veterans' Preference on the Composition and Quality of the Federal Civil Service," *Journal of Public Administration Research and Theory*, Vol. 23, No. 2, 2013, pp. 247–265.

Lim, N., L. T. Mariano, A. G. Cox, D. Schulker, and L. M. Hanser, *Improving Demographic Diversity in the U.S. Air Force Officer Corps,* Santa Monica, Calif.: RAND Corporation, RR-495-AF, 2014. As of August 12, 2015: http://www.rand.org/pubs/research_reports/RR495.html

Lopez, C. T., "Recruiting Force Remains Unchanged, Despite Shrinking Goals," Army News Service, January 16, 2014. As of August 12, 2015: http://www.ctlopez.com/story/?story=20140116_3887

Losey, S., "Air Force Announces Rollbacks to Speed Separations," *Air Force Times*, December 17, 2013a. As of August 12, 2015: http://www.airforcetimes.com/article/20131217/NEWS07/312170022/ Air-Force-announces-rollbacks-speed-separations

———, "Airmen Facing Quality Force Review Board to Be Notified Jan. 6," *Air Force Times*, December 23, 2013b. As of August 17, 2015: http://www.airforcetimes.com/article/20131223/NEWS/312230013/ Airmen-facing-Quality-Force-Review-Board-notified-Jan-6

———, "Board Targets Enlisteds in Overmanned Fields for Cuts," *Air Force Times*, December 30, 2013c. As of August 17, 2015: http://www.airforcetimes.com/article/20131230/CAREERS02/312300017/ Board-targets-enlisteds-overmanned-fields-cuts

———, "Cash to Leave in 2015: AF Cranks Up Retirement, Separation Budget," *Air Force Times*, March 24, 2014a. As of August 17, 2015: http://www.airforcetimes.com/article/20140324/CAREERS/303240029/ Cash-leave-2015-AF-cranks-up-retirement-separation-budget

———, "3,500 to Be Ousted by Quality Force Review Board," *Air Force Times*, June 5, 2014b. As of August 17, 2015: http://www.airforcetimes.com/article/20140605/ NEWS/306050086/3-500-ousted-by-quality-force-review-board

McCormick, D., *The Downsized Warrior: America's Army in Transition*, New York: New York University Press, 1998.

Mehay, S. L., and P. F. Hogan, "The Effect of Bonuses on Voluntary Quits: Evidence from the Military's Downsizing," *Southern Economic Journal*, Vol. 65, 1998, pp. 127–139.

Military Leadership Diversity Commission, *Compelling Government Interests and Diversity Policy*, MLDC Issue Paper No. 36, Arlington, Va., May 2010a.

———, *An Overview of Civil Cases Challenging Equal Opportunity Guidance to Certain Military Promotion and Retirement Boards*, MLDC Issue Paper No. 51, Arlington, Va., November 2010b.

———, *Retention*, MLDC Decision Paper No. 3, Arlington, Va., February 2011a.

———, *From Representation to Inclusion: Diversity Leadership for the 21st-Century Military*, Final Report, Arlington, Va., 2011b.

Miller, B. D., *The Impact of the Drawdown on Minority Officer Retention*, thesis, Monterey, Calif.: Naval Postgraduate School, 1995.

Miller, L. L., J. Kavanaugh, M. C. Lytell, K. Jennings, and C. Martin, *The Extent of Restrictions on the Service of Active-Component Military Women*, Santa Monica, Calif.: RAND Corporation, MG-1175-OSD, 2012. As of August 17, 2015: http://www.rand.org/pubs/monographs/MG1175.html

MLDC—*See* Military Leadership Diversity Commission.

Nataraj, S., L. M. Hanser, F. Camm, and J. Yeats, *The Future of the Army's Civilian Workforce: Comparing Projected Inventory with Anticipated Requirements and Estimating Cost Under Different Personnel Policies*, Santa Monica, Calif.: RAND Corporation, RR-576-A, 2014. As of August 17, 2015: http://www.rand.org/pubs/research_reports/RR576.html

Odierno, R. T., *Planning for Sequestration in Fiscal Year 2014 and Perspectives of the Military Services on the Strategic Choices and Management Review*, Statement Before the House Armed Services Committee, First Session, 113th Congress, September 18, 2013.

Office of the Deputy Under Secretary of Defense (Military Community and Family Policy), *2004 Demographics: Profile of the Military Community*, 2004. As of August 17, 2015: http://www.militaryonesource.mil/12038/MOS/Reports/Combined%20Final%20Demographics%20Report.pdf

Quester, A. O., and C. L. Gilroy, *America's Military: A Coat of Many Colors*, Alexandria, Va.: CNA Corporation, 2001.

Parcell, A., "Officer Force-Shaping Tools: Voluntary Programs and Tradeoffs with Other Force-Shaping Goals," memorandum to CAPT Bradley Mai, Navy BUPERS-31, Alexandria, Va.: CNA, CME D0024494.A1/2, February 2011.

Public Law 113-66, National Defense Authorization Act for Fiscal Year 2014, Sec. 904, Streamlining of Department of Defense Management Headquarters, Washington, D.C.: U.S. Government Printing Office, December 26, 2013.

Reilly, C., "Navy Thinning Is Forcing out Thousands of Sailors," *Virginian-Pilot*, March 2, 2012. As of August 17, 2015: http://hamptonroads.com/2012/03/navy-thinning-forcing-out-thousands-sailors

Reilly, S., "RIFs Pose Tough HR Challenge: Personnel Cuts Likely at DoD," *Federal Times*, October 28, 2013. As of August 17, 2015: http://archive.federaltimes.com/article/20131028/PERSONNEL/310280010/RIFs-pose-tough-HR-challenge

Roth, P. L., P. Bobko, and F. S. Switzer, III, "Modeling the Behavior of the 4/5ths Rule for Determining Adverse Impact: Reasons for Caution," *Journal of Applied Psychology*, Vol. 91, No. 3, May 2006, pp. 507–522.

Rostker, B., *I Want You! The Evolution of the All-Volunteer Force*, Santa Monica, Calif.: RAND Corporation, MG-265-RC, 2006.

———, *Right-Sizing the Force: Lessons for the Current Drawdown of American Military Personnel*, working paper, Washington, D.C.: Center for a New American Security, June 2013.

Sanborn, J. K., "'Surplus' Officers: Retire Early or Risk SERB," *Marine Corps Times*, August 6, 2012. As of August 17, 2015:
http://www.marinecorpstimes.com/article/20120806/
NEWS/208060319/-8216-Surplus-officers-Retire-early-risk-SERB

Schneier, C., "Hagel Warns of Possible DoD Layoffs in 2014 if Sequestration Continues," *Federal News Radio*, July 10, 2013. As of August 17, 2015:
http://www.federalnewsradio.com/394/3385257/
Hagel-warns-of-possible-DoD-layoffs-in-2014-if-sequestration-continues

Schroetel, A. H., "Military Personnel: End Strength, Separations, Transition Programs and Downsizing Strategy," Annex J to *Adjusting to the Drawdown: Report of the Defense Conversion Commission*, Washington, D.C., 1993.

Schuette v. Coalition to Defend Affirmative Action, 572 U.S. __ (2014), Docket No. 12-682.

Schwellenbach, N., "Are Pentagon Civilians Really Behind the Pentagon's Money Woes?" *Time*, June 4, 2013. As of August 17, 2015:
http://nation.time.com/2013/06/04/
are-pentagon-civilians-really-behind-the-pentagons-money-woes/

Secretary of the Air Force Public Affairs, "June 19—Pulse on AF Force Management," U.S. Air Force News Service, June 20, 2014a. As of August 17, 2015:
http://www.af.mil/News/ArticleDisplay/tabid/223/Article/485692/june-19-pulse-on-af-force-management.aspx

———, "Aug. 1—Pulse on AF Force Management," U.S. Air Force News Service, August 1, 2014b. As of August 17, 2015:
http://www.af.mil/News/ArticleDisplay/tabid/223/Article/490623/aug-1-pulse-on-af-force-management.aspx

Tan, M., "Army Launches New Incentives to Quit Active, Join Reserves: Guard and Reserve Test New Program for Drawing Active-Duty Soldiers," *Army Times*, February 17, 2014. As of August 17, 2015:
http://www.armytimes.com/article/20140217/NEWS/302170006/
Army-launches-new-incentives-quit-active-join-reserves

Under Secretary of Defense for Personnel and Readiness, "Civilian Workforce Shaping," memorandum, Washington, D.C., September 12, 2013.

U.S. Air Force, Air Force Instruction (AFI) 36-2107, *Active Duty Service Commitments*, Washington, D.C., November 25, 2009.

U.S. Air Force Audit Agency, *Air Force Personnel Reductions: Audit Report*, F2008-004-FD4000, May 12, 2008. As of August 20, 2015: http://www.foia.af.mil/shared/media/document/AFD-100528-099.pdf

U.S. Department of Defense, *Department of Defense Plan for Streamlining the Bureaucracy*, 797, December 1993. As of August 20, 2015: https://archive.org/details/DepartmentofDefensePlanforStreamliningtheBureaucracy

U.S. Department of Defense, *The Defense Science Board Task Force on Human Resources Strategy*, Washington, D.C.: Office of the Under Secretary of Defense for Acquisition, Technology, and Logistics, February 2000.

———, *Population Representation in the Military Services, Fiscal Year 2011*, Washington, D.C.: Office of the Under Secretary for Personnel and Readiness, 2011. As of August 18, 2015: http://prhome.defense.gov/RFM/MPP/AP/POPREP.aspx

———, Department of Defense Directive (DoDD) 1020.02, *Diversity Management and Equal Opportunity (EO) in the Department of Defense*, Washington, D.C.: Office of the Under Secretary of Defense for Personnel and Readiness, February 5, 2009.

———, Department of Defense Instruction (DoDI) 1332.32, *Selective Early Retirement of Officers on an Active Duty List and the Reserve Active Status List and Selective Early Removal of Officers from the Reserve Active Status List*, Washington, D.C.: Office of the Under Secretary of Defense for Personnel and Readiness, December 27, 2006.

———, Department of Defense Instruction (DoDI) 1400.25, Vol. 1702, *DoD Civilian Personnel Management System: Voluntary Separation Programs*, Washington, D.C.: Office of the Under Secretary of Defense for Personnel and Readiness, April 1, 2009.

———, Department of Defense Instruction (DoDI) 1332.30, *Separation of Regular and Reserve Commissioned Officers*, Washington, D.C.: Office of the Under Secretary of Defense for Personnel and Readiness, September 20, 2011.

———, *Defense Budget Priorities and Choices—Fiscal Year 2014*, April 2013.

U.S. Department of the Navy, Office of the Chief of Naval Operations, OPNAV Instruction 1811.3A, *Voluntary Retirement and Transfer to the Fleet Reserve of Members of the Navy Serving on Active Duty*, Washington, D.C., February 28, 2012.

U.S. General Accounting Office, *Military Personnel: High Aggregate Personnel Levels Maintained Throughout Drawdown*, Washington, D.C., GAO/NSIAD-95-97, June 1995.

————, *Army Force Structure: Future Reserve Roles Shaped by New Strategy, Base Force Mandates, and Gulf War*, Washington, D.C., GAO/NSIAD-93-80, December 15, 1992.

————, *Defense Civilian Downsizing: Challenges Remain Even with Availability of Financial Separation Incentives*, Washington, D.C., GAO/NSIAD-93-194, May 14, 1993.

————, *Civilian Downsizing: Unit Readiness Not Adversely Affected, but Future Reductions a Concern*, Washington, D.C., GAO/NSIAD-96-143BR, April 22, 1996.

————, *Force Structure: Potential Exists to Further Reduce Active Air Force Personnel*, Washington, D.C, GAO/NSIAD-97-78, March 28, 1997.

————, *Quadrennial Defense Review: Some Personnel Cuts and Associated Savings May Not Be Achieved*, Washington, D.C., GAO/NSIAD-98-100, April 30, 1998.

U.S. Government Accountability Office, *Force Structure: Assessments of Navy Reserve Manpower Requirements Need to Consider Most Cost-Effective Mix of Active and Reserve Manpower to Meet Mission Needs*, Washington, D.C., GAO-06-125, October 18, 2005.

U.S. Marine Corps, *Reshaping America's Expeditionary Force in Readiness*, Report of the 2010 Marine Corps Force Structure Review Group, Washington, D.C., Department of the Navy, March 14, 2011.

————, *Convening of the FY13 Selective Early Retirement Board (SERB) to Recommend Regular Unrestricted Lieutenant Colonels on the Active-Duty List of the Marine Corps for Selective Early Retirement*, MARADMIN 419/12, Washington, D.C., August 1, 2012. As of August 18, 2015:
http://www.marines.mil/News/Messages/MessagesDisplay/tabid/13286/
Article/110447/
convening-of-the-fy13-selective-early-retirement-board-serb-to-recommend-regula.
aspx

U.S. Navy, *FY12 Active Component Unrestricted Line Captain and Commander Selective Early Retirement Board*, NAVADMIN 006/11, Washington, D.C., January 2011. As of August 18, 2015:
http://www.public.navy.mil/bupers-npc/reference/messages/Documents/
NAVADMINS/NAV2011/NAV11006.txt

U.S. Navy Reserve, "Navy Reserve Component Force Changes and Manning Actions Frequently Asked Questions," November 29, 2012. As of August 18, 2015:
http://88.198.249.35/preview/hfeUqxjG_CSTt31IzL3OoD8nhi-
s9JhBkqNYjazfRtU,/Navy-Reserve-Component-Force-Changes-and.
html?query=NOSC-Baltimore-Website

Vanden Brook, T., "Black Officers Dismissed at Greater Rate than Whites," *USA Today*, August 4, 2014. As of August 12, 2015:
http://www.usatoday.com/story/news/nation/2014/08/04/
army-black-majors-dismissed-higher-rates-than-whites/13587821/

Walker, D. M., *Human Capital: DOD's Civilian Personnel Strategic Management and the Proposed National Security Personnel System*, Testimony Before the Subcommittee on Oversight of Government Management, the Federal Workforce and the District of Columbia, Senate Committee on Governmental Affairs, Washington, D.C.: U.S. General Accounting Office, GAO-03-493T, May 2003.

Wiggins, A., "Qualitative Service Program Selects NCOs for Separation, Retention," U.S. Army website, November 29, 2012. As of January 15, 2014:
http://www.army.mil/article/92042

Wyatt, J. R., *Navy Recruiting and Retention: Yesterday, Today and Tomorrow*, research report, Maxwell Air Force Base: Air Command and Staff College, Air University, No. AU/ACSC/234/1999-04, 1999.

Zedeck, S., "Adverse Impact: History and Evolution," in J. L. Outtz, ed., *Adverse Impact: Implications for Organizational Staffing and High Stakes Selection*, New York: Routledge, 2009, pp. 3–28.